THE PAINTED STORK

THE PAINTED STORK

STORK

Exploring Ecology and
Conservation in India

ABDUL JAMIL URFI

PELAGIC PUBLISHING

First published in 2024 by
Pelagic Publishing
20–22 Wenlock Road
London N1 7GU, UK

www.pelagicpublishing.com

The Painted Stork: Exploring Ecology and Conservation in India

https://doi.org/10.53061/NIOF6501

British Library Cataloguing in Publication Data
A catalogue record for this book is available from the British Library

ISBN 978-1-78427-439-9 Pbk
ISBN 978-1-78427-440-5 ePub
ISBN 978-1-78427-441-2 PDF

Cover photo: A Painted Stork carrying nesting material. © N. K. Tiwary

Frontispiece: N. K. Tiwary

Printed and bound in Great Britain by
TJ Books Limited, Padstow, Cornwall

For

Zara

Contents

Foreword

I know the author from his intensive and detailed field studies of Oystercatchers, which I also used to study, and the research papers he produced. His lifetime's passion, though, is the Painted Stork, about which he has written numerous research papers, popular articles and a previous monograph. This book brings together his work and knowledge in a very readable manner.

While this enjoyable volume concentrates on Painted Storks, it also reviews (including terrific photographs) all the stork species and a range of other colonial birds.

Painted Storks are an excellent species for study as they are conspicuous and easily monitored; the colony at Delhi Zoo provides a perfect location for the author's core studies. Combining theoretical concepts, detailed fieldwork and observations from across the species' range, this wide-ranging book covers all aspects of the Painted Stork – from mythology to history to biology to adaptation in a changing world.

Conservation permeates the book, whether in the context of adaptation to urban environments, the role of protection, invasive species or climate change (especially the key relationship with the monsoon). This is just the sort of work we need to ensure our waterbirds have a healthy future.

William J. Sutherland
Professor of Conservation Biology
Department of Zoology
University of Cambridge
2023

Prologue

Some of us fascinated by this planet's magnificent biodiversity face the dilemma of choosing one out of the so many organisms on offer to adopt as our life-long subject of study. In *To Know A Fly*, a little 1962 classic filled with wisdom, humour and artful prose, Vincent Dethier made the following prescription: 'The answer is simple', he said. 'Let the species choose you.' But having selected a species or being chosen by one, how does one go about studying it? In his inspiring 1994 memoir *Naturalist*, the reading of which I have described elsewhere as 'the most pleasurable way to learn, reflect and shape one's career in science', E.O. Wilson remarks: 'Love the organisms for themselves first, then strain for general explanations, and, with good fortune, discoveries will follow. If they don't, the love and the pleasure will have been enough.'

With or without reading Dethier and Wilson, Abdul Jamil Urfi appears to have heeded both pieces of advice, and with much success. Urfi's is an all too familiar story. As a young boy, he was attracted by the outdoors and fascinated by animals. It should be logical, should it not, that he would choose to study zoology? He did, but the zoology taught in the classroom was worse than uninspiring. Urfi describes his experience in moving terms: 'My chief regret was that few people seemed interested in zoology itself, especially behaviour and ecology which appealed to me. The [Zoology Department] buildings seemed like molluscan shells piled on the seashore – looking beautiful from the outside, with their intricate carvings and patterns. But the animal that once inhabited them had died long ago. Alas, the truth was that nobody was interested in zoology.'

I know many a wildlife enthusiast whose love of animals was irreversibly extinguished by studying zoology! But Urfi is made of sterner stuff. He escaped to the outdoors at every opportunity and, in his own words, 'discovered zoology through birdwatching'. Echoing India's most famous Ornithologist Salim Ali's sentiment that 'Birdwatching is like measles. You have to catch the disease', he reflects: 'I had caught that disease long ago, and when I began to tire of the dull and boring indoor lectures and practicals in the Zoology Department, it came to my rescue.'

One of the many outdoor places Urfi hung out was Delhi Zoo. Just as he was more interested in looking outside the classrooms of the zoology department, so was he more interested in looking outside the animal enclosures at the zoo. It turned out that a wild colony of the Painted Stork *Mycteria leucocephala* had been nesting every year in the trees planted on the little islands on the zoo premises. What a wonderful use of this space, a true interpretation of the 'zoological park' epithet! As if that were not enough, the then Director of Delhi Zoo, J.H. Desai, had also showered his benevolent attention on the wild birds 'encroaching' his territory. Desai published a detailed account of the zoo's Painted Stork population, prompting Urfi to pay tribute with Issac Newton's quote, 'If I have seen further, it is only by standing on the shoulders of giants.'

It is here that the Painted Stork appears to have chosen Abdul Jamil Urfi as its chief documenter and spokesman. Over the next 35 years, Urfi researched this and other populations of this species with the love for the organism prescribed by Wilson – and he has not failed to strain for general explanations. And as Wilson predicted, success has followed. Conducting field studies in Delhi, Rajasthan and Karnataka, Urfi and his students have broken much new ground in investigating the foraging ecology, nesting habits, evolution of sexual size dimorphism, mating patterns, coloniality and genetic diversity in the Painted Stork.

Urfi synthesised all his discoveries and placed them in the context of the international literature on this and related species in his 2011 monograph *The Painted Stork: Ecology and Conservation*. That book was aimed at a scientific scholarly audience. The present work is a more personal and accessible account of his research, aimed especially at students who may be developing a similar fascination for some particular species. An equally important goal of this book is to encourage students to engage with the problems and prospects of nature conservation, not merely as ideological or moral commitments but as a scientific endeavour.

This book covers much territory, introducing the Painted Stork and its relatives in a zoological context, providing interesting snippets about its perception and depiction in folklore around the world, and discussing many aspects of its behaviour and ecology – all broadly mapping onto Urfi's research interests. The three final chapters on the impact of urbanization, the role of local peoples in shaping this bird's survival and distribution and, finally, complex issues of conservation, are especially valuable. Furthermore, their worth goes beyond solely the conservation of the Painted Stork, as they contain general lessons for any conservation effort in the face of the inevitable forces of development, invasive species, the complex dynamics of monsoons and ongoing climate change. Coming at the end of a detailed study of the behaviour and ecology of a single species, these three chapters showcase the critical role of science in conservation.

I am pleased to recommend *The Painted Stork: Exploring Ecology and Conservation in India* to a wide range of readers all over the world and especially in India. There is a great dearth of role-models and books that inspire and inform us about the potential for conducting first-rate science that can be combined with a passion for the outdoors, a love of wildlife, the spirit of adventure and freedom from the need to procure large grants and laboratory facilities. Urfi's life and work fulfil this need admirably and have the power to produce a new generation of role-models and books that carry forward the spirit of science embodied in them.

Raghavendra Gadagkar
Centre for Ecological Sciences, Indian Institute of Science
September 2022

Preface

I must confess that this volume came to be written quite accidentally. I had originally planned a book along the lines of a personal narrative of my lifelong involvement with studying birds, first the European Oystercatcher *Haematopus ostralegus* while a student in England, and later the Painted Stork *Mycteria leucocephala*, for much of my career as a teacher and researcher in India. That book, I had hoped, would be peppered with anecdotes and personal experiences and I would attempt to provide the reader with a flavour of what it is like to study birds and embark upon a career as a field biologist in India. I soon discovered, however, that my stories of learning how to do science had little market in today's competitive publishing scenario. (And, anyway, I wrote up some of my experiences as articles, which are available in the public domain.) So, when it was suggested that a book on the Painted Stork might work as a popular monograph to acquaint an international readership – especially students of ecology and conservation – with this fascinating species, I jumped at the idea.

This book was written with the aim of showing how birds can serve as tools in conservation biology, and how in the context of India more studies of birds could be initiated in order to address issues of ecology and conservation. One of my central objectives behind attempting to write such a book as this was environmental education, especially to facilitate an appreciation of ecology and conservation in a broader context. I have tried to touch upon some very basic issues, even though some of them may seem intuitive and obvious. In this context, my experience of working with the Centre for Environment Education (CEE) in Ahmedabad – a centre for excellence in environmental education – came in very handy. The five years I spent there as a member of staff demonstrated to me the value of communicating for the environment, besides giving me ample opportunities to carry out my studies on the Painted Stork in the state of Gujarat (Western India). Finally, a review of all studies on the Painted Stork resulted in my book, a species monograph entitled *The Painted Stork: Ecology and Conservation*, published about a decade ago. The present book is essentially an updated, revised and more illustrated and version of the earlier book, to which I refer extensively.

October 2023

Acknowledgements

During my journey of 35-plus years on the trail of the Painted Stork, several people – teachers, colleagues and students – have provided help, support and encouragement at various stages. Most of them have been acknowledged in my previous book on this subject. As far as this book is concerned I would again like to thank some of the people who made an important difference. I thank Prof. C. R. Babu, Prof. T. R. Rao (both from the University of Delhi) and Prof. L. K. Pande (Jawaharlal Nehru University, Delhi) for their support and encouragement and Kartikeye V. Sarabhai, the Director of CEE for encouraging my studies on the Painted Stork in Gujarat. I have profited immensely from my association with Dr J. D. Goss-Custard, my supervisor when I was a student learning behavioural ecology and field ecology research by working on the Oystercatchers of the Exe estuary. I thank Prof. Peter Frederick, my host at the University of Florida–Gainesville, who opened the doors for research on the Wood Stork *Mycteria americana* and the ecology of the Everglades while I was a Fulbright scholar in the United States. I thank all my students at the University of Delhi who completed their MSc projects and PhD dissertations under me. I thank my fellow traveller on the 'wading birds foraging ecology trail', Prof. R. Nagarajan, for his help and my friends Rebecca Spurk, Kandarp Kathju, Anindya Sinha, Amitabh Joshi and Zaheer Baber for their support. I thank Prof. Bill Sutherland for writing the foreword and Prof. Raghavendra Gadagkar for writing the prologue. A number of people helped in preparation of this book, some by providing photographs from their personal collections and others by helping in preparation of illustrations. I am particularly grateful to the following for their help: Dr Asad Rafi Rahmani, Dr Mahendiran Mlyswamy, Dr Bharat Bhushan Sharma, Dr Nawin K. Tiwary, Mr. Paritosh Ahmed and Mr. Sohail Akbar. I wish to thank Rahul Rohitashwa for sending me pictures of Painted Storks on first day covers and postage stamps. I thank Dr Pratibha Baveja for bringing me up to date about the issue of hybridisation between Painted Stork and Milky Stork in South-East Asia and sharing her work done at the National University of Singapore. Thanks also to Mr Nahar Singh, Director of KDGNP, for permitting me to take pictures of dioramas (of heronries) at the Salim Ali Visitor Interpretation Centre, Bharatpur. I am grateful to Nigel Massen, David Hawkins and Sarah Stott at Pelagic for all their cooperation and support.

Painted Stork gaze. (N. K. Tiwary)

Introduction

A Bird of Great Charisma

The Painted Stork is a large, colonially nesting wading bird with a prominent yellow bill. Listed as near threatened (NT) by international conservation agencies (BirdLife International 2023a), it is found across large parts of South Asia and South-East Asia, with a stronghold on the Indian subcontinent, particularly in India and Sri Lanka. The genus *Mycteria*, to which it belongs, has representatives in three continents – Asia, America and Africa – each differing from the other in minor details.

My own association with the Painted Stork, spanning more than 35 years, has been influenced by a host of personal and professional factors. The interesting point, though, is that for studying these attractive birds I did not have to venture very far most of the time. In my hometown of Delhi – India's capital city – a nesting colony of Painted Storks has been in existence since 1960. Each year at the end August or in early September, these birds start congregating within the premises of the city's zoo. Here they build their nests on trees planted on islands in its ponds. During this period, broadly from August/September to March, they raise their young, oversee preliminary rites of passage and then, as soon as spring is in the air and the winter is taking its last gasps, they are all gone. For the rest of the year, they remain widely dispersed in the surrounding countryside, where they live singly or in loose scattered groups, seeking food and shelter. Over the years I have seen this pattern repeat itself again and again (Urfi 2011a).

While I have focused much of my research effort on the colony at National Zoological Park Delhi (hereinafter 'Delhi Zoo'), I have also studied the Painted Stork and other species of colonially nesting waterbirds at other places in India. But the Delhi Zoo colony is very special and has several advantages when it comes to considering various aspects of Painted Stork biology and behaviour. First of all, it is easy and convenient for studying these birds at close quarters. Painted Storks build their nests on trees in concrete-lined ponds that are in view from pathways located just a few metres away. I should emphasise that although located within the premises of a zoo, this population is completely free ranging and wild. This Painted Stork colony has been well studied, not just by me and my group at the University of Delhi, but by others too. There is a large quantity of work on different facets of Painted Stork biology already available. I daresay that, due to our collective efforts, this population is perhaps the best studied of any population of wild birds in India (Figure I.1).

Why study the Painted Stork?

What good can come from a study of the Painted Stork? Why indeed should it be studied, besides the fact that it is there, within easy reach? I think there are several strong reasons why a study of this wetland bird can be meaningful. Large birds are always an apposite focus for ecological studies simply because they can be easily seen and counted and help us to understand a number of important issues in ecology and conservation. The Painted Stork is an indicator of its habitat, wetlands, which themselves merit attention due to their threatened status. Across much of Asia, the Indian subcontinent in

FIGURE I.1 A Painted Stork nesting within the precincts of the Delhi Zoo, with the ramparts of the Old Fort visible in the background. (N. K. Tiwary)

particular, wetlands are being lost to land encroachment, pollution and other factors. Yet several crucial aspects of their ecology, particularly the role of birds in ecosystem functioning, remain largely unexplored. Since environmental toxins, particularly organochlorine pesticides, travel rapidly along aquatic food chains, the study of piscivorous birds like the Painted Stork assumes a special significance. (At each trophic level the concentrations of pesticides increases in the bodies of organisms by a process known as 'biomagnification'.)

The Painted Stork provides an opportunity to explore questions about avian coloniality, a topic that has fascinated ornithologists and evolutionary biologists for a very long time. While the literature on this subject is extensive, detailed studies on colonial non-passerines, and particularly heronry birds, are few in number and it is here that studies on birds like the Painted Stork can be extremely useful. The fact that this highly visible and easily studied bird nests all across the country – its breeding colonies in northern and southern India separated by nearly 20° of latitude – makes it an excellent model to study local bioclimate-driven ecomorphological changes in wild birds and to explore interesting questions in biogeography.

While wetlands – the foraging habitat of the Painted Stork and other fish-eating birds – are gravely threatened, equally vulnerable today are the nesting colonies referred to as heronries. In India, heronries are not located just in marshes; many are associated with village reservoirs and the wider countryside, inside and outside the protected areas network and also in parks and gardens in urban settings, too. The Delhi Zoo colony is a good example of a heronry in an urban setting. Perhaps because their natural nesting areas are getting scarce in the countryside due to the cutting down of trees and habitat loss, colonial waterbirds look for suitable nesting sites safe from human disturbance in urban parks and gardens. Be that as it may, these birds serve as indicators of the changes going on in their environment: namely, the spread of urbanisation and the increasing extent of built-up areas. Monitoring these birds can help in keeping tabs on the environment and can therefore yield interesting insights for city planning and development. However, what parameters of the population or individuals to monitor, and how to monitor them, is the turf of biologists, as I will elaborate on later.

Birds are good models for monitoring effects of global climate change (Pearce-Higgins and Green 2014). One of the most significant dimensions of the study of the Painted Stork is their dependence upon the monsoon. How exactly do these seasonal rains govern the food cycles in wetlands? What happens when the monsoon fails? Several years ago, when I was beginning to get interested in the Painted Stork, I thought deeply about how the monsoon impacts biodiversity and our lives. Although I did not have a full understanding of all the issues involved, I ended up writing a well-received article in *Natural History* magazine (Urfi 1998, Figure I.2), which examined the link between the Painted Stork and the monsoon rains, which provide the bulk of the annual

FIGURE I.2 Cover of the *Natural History* magazine issue which carried the story on Painted Stork 'A monsoon delivers stork'. (From *Natural History*, November 1998, © Natural History Magazine, Inc., 1998)

rainfall in countries across Asia and are regarded as the lifeline of these nations. Much of Asia, which has a predominantly monsoonal climate, experiences two phases: a dry phase when the countryside looks grey and forbidding, and a wet phase during the rains when everything is fresh and green. This is the time when wetlands are rejuvenated and aquifers recharged. Most places in India receive the bulk of their annual rainfall (in some places close to 80%) during the three or four months of the monsoon. Though there are fears that the monsoon regime is being impacted by global climate change (of which there are tell-tale signs already) no organism has yet been identified as a good model for studying the effects of seasonal rains on biodiversity. The food supply of fish-eating birds is strongly influenced by the monsoon, so studying species such as the Painted Stork may help to fill gaps in understanding.

About this book

Using research on the Painted Stork I hope to explain issues related to ecology and conservation more generally, in a manner that will be useful for researchers and students at all levels, in colleges and universities across Europe, Asia and other parts of the world, as well as to naturalists and scientists. Since the subjects under discussion are diverse – ornithology, ecology, evolutionary biology, the human–wildlife interface and urbanisation, amongst others – I take a holistic view and attempt to integrate them all, with studies on the Painted Stork being the binding thread. I frequently use the comparative approach by using studies on other species, particularly the Wood Stork, to explain a point. Specific topics are explored in boxes alongside the main text. While suggesting that a study of the Painted Stork and similar birds can yield meaningful insights into the problems of urbanisation, climate change and so on, I do not mean rigorous research exists, undertaken either by my own group at the University of Delhi or others, which has already established the effects or problems raised in context. My attempt here is to provide the necessary ecological background for understanding some of these issues and how studies on birds like the Painted Stork can be meaningful in future.

Structure of the book

Chapter 1, The Painted Stork in Context, starts with an outline of myths and stories concerning storks, largely from Western tradition. The bulk of the chapter offers an overview of storks worldwide and ends with a discussion on the genus *Mycteria*, to which the Painted Stork belongs. A glimpse into the contributions of some prominent workers on storks is also provided, as well as an account of the representation of heronry birds in Indian art. A mention is made about recently published studies on hybridisation between the Painted Stork and Milky Stork in South-East Asia.

Chapter 2, Avian Coloniality. Since one of the defining features of the Painted Stork is its colonial nesting habit, this chapter provides a general overview of avian coloniality. An attempt is made to familiarise the reader with the broad contours of the topic. A comprehensive and updated review of avian coloniality is not attempted here; however, there is a discussion on some of the approaches that ecologists have used to understand this form of nesting behaviour. An explanation of some commonly used terms and concepts is included in this chapter.

Chapter 3, Painted Stork Colonies in India considers how these birds nest in mixed-species colonies all across India, both inside and outside the protected areas network, in rural and urban areas. Some of the well-known Painted Stork nesting colonies or large mixed heronries are described. Their geographical location sets the stage for a discussion on biogeography and nesting period in relation to food availability, discussed in detail in Chapter 6.

Chapter 4, Nesting presents current understanding of nesting and explains how results from recent field studies have been used to model nesting success in the Painted Stork population of Delhi Zoo.

Chapter 5, Sexual Size Dimorphism deals with studies undertaken to improve understanding of sexual size dimorphism by using the innovative, non-invasive technique of videography of the Delhi Zoo population of Painted Stork. I take stock of some recent studies on the morphometrics of the species across India undertaken by using similar non-invasive methods (photogrammetry).

Chapter 6, Foraging Ecology. The Painted Stork is a wetland bird and its nesting is strongly related to the hydrological cycles and their effects on food availability. This chapter sets out the basis for a discussion of the importance of the seasonal monsoon rains in the food cycles of wetlands. Details of recent studies on the foraging ecology of the Painted Stork are also presented.

Chapter 7, Painted Storks in an Urban Context. As Painted Storks and other colonial waterbirds nest regularly in several Indian cities, the issue of heronries in urban areas assumes increasing significance, especially as natural nesting habitats are coming under growing pressure in the countryside. The fluctuations in populations of colonial waterbirds in urban parks and their adaptations are analysed in this chapter.

Chapter 8, Painted Storks and People. This chapter examines the views of different stakeholders in the conservation of stork nesting colonies in the countryside and how people look upon these birds. A special focus on Kokkare Bellur village where Painted Storks and Spot-billed Pelican *Pelecanus philippensis* breed in large numbers is included.

Chapter 9, Conservation, discusses the importance of monitoring programmes and also takes stock of various initiatives by conservation activists and organisations as they attempt to safeguard the future of the species.

Painted Stork in soaring flight. (N. K. Tiwary)

Chapter 1

The Painted Stork in Context

In public perception storks are regarded as being connected with the birth of babies. Storks are often depicted on greeting cards carrying a baby wrapped in a bundle in their bill. Philip Kahl, a leading stork researcher whose work we will discuss later, quipped about this largely Western tradition, 'a cute and convenient way to avoid early sex education' (Kahl 1971b). In many European countries, putting a model of a White Stork *Ciconia ciconia* with a bundle in its bill in front of the house where a baby has been recently born is a popular custom (Figure 1.1).

The fact that storks are large birds that are regularly seen in certain localities and possess interesting behavioural traits means that they tend to be readily noticed; this perhaps explains why they figure quite prominently in folklore and cultural traditions in many countries, such as the presence of the White Stork in Aesop's fables. In Western traditions, storks mostly have positive sentiments associated with them. In many places in Europe they are regarded as the herald of spring and therefore linked to rebirth, as well as purity and piety. This tall stately bird is well regarded not just in its breeding grounds in parts of north-eastern and southern Europe, but also across considerable parts of its wintering range, particularly in North Africa and Asia.

One of the most prominent features of storks is their long, strong bill, something that lends itself easily to symbolism of the phallus and, in consequence, fertility. Because of their quiet, industrious nesting activity exhibited by a breeding pair and the focus of their activities on rearing chicks and raising a family, storks also symbolise domestic concerns and values. One may consider that storks are not 'sexy' and lack the panache of some other birds, such as birds of prey or pheasants, with their bold, brilliant plumage.

FIGURE 1.1 A model of a stork displayed outside a home in Saarbruicken, south-
western Germany, where a baby had been recently born. (Rebecca Spurk)

Storks worldwide

The Painted Stork belongs to the ornithological order Ciconiiformes and the
family Ciconiidae, which has 19 species recognised worldwide (Sibley and
Ahlquist 1990). Standard ornithological texts such as del Hoyo et al. (1992)

treat family Ciconiidae as comprising three distinct tribes, Mycterini, Ciconiini and Leptoptilini. Besides the Painted Stork, a brief description of the other species of stork in India is given below. The abbreviations used for the conservation status of each species are as follows: LC = Least Concern, NT = Near Threatened, VU = Vulnerable and EN = Endangered. Further information about all these species is available online (e.g., Birds of the World 2023).

Mycterini comprises the wood storks and openbills and includes the genera *Mycteria* and *Anastomus.* The latter has two species: Asian Openbill *Anastomus oscitans* and African Openbill *A. lamelligerus.* As their common name suggests, these storks have an opening or a gap between the mandibles which somehow assists with the capture of snails (molluscs), their principal prey. All the species in this tribe are typified by their smaller size, colonial nesting habit, and a high degree of specialisation in their feeding techniques and trophic apparatus.

Asian Openbill or **Asian Openbill Stork** *Anastomus oscitans* LC (Figure 1.2). This is a comparatively small stork, measuring 81 cm in length and standing 68 cm tall. It is chiefly greyish-white (non-breeding season) or white (breeding season), with glossy black wings and a tail that has a green or purple sheen. Its name is derived from the distinctive gap formed between the recurved lower

FIGURE 1.2 Asian Openbill Stork. (N. K. Tiwary)

and arched upper mandible of the bill in adult birds. Young birds do not have this gap. The cutting edges of the mandible have a fine brush-like structure that is thought to give them better grip on the shells of snails. Its short legs are pinkish to grey, but reddish prior to breeding. Non-breeding birds have smoky-grey wings. Young birds are brownish-grey and have a brownish mantle. They are usually found in flocks but single birds are not uncommon. Like all storks, it flies with its neck outstretched. The Asian Openbill occurs in Bangladesh, Cambodia, India, Myanmar, Nepal, Pakistan, Sri Lanka, Thailand and Vietnam, and Laos; it is also recorded from Bhutan though its breeding is uncertain there. Although resident within their range, they make long-distance movements in response to weather and food availability. Nestlings ringed at Bharatpur have been recovered up to 800 km due east (Urfi 2011a). Like other storks, the Asian Openbill is a broad-winged, soaring bird, which relies on sailing on thermal currents for long-distance flights.

The Asian Openbill feeds mainly on large molluscs, especially *Pila* species. The relationship between the monsoon rains and food cycles of the Asian Openbill remain to be studied in detail, though it is held that the first monsoon showers trigger the movement of molluscs buried in the ground. The exact manner in which the bird uses the gap between its mandibles has been a matter of speculation that well-known biologists (for instance, Sir Julian Huxley) have waded into. However, the Asian Openbill is also known to feed on water snakes, frogs and large insects. It may be seen in inland waters, such as *jheels* and marshes, but is rarely observed in rivers or on tidal mudflats. Like other storks, this species is silent except for occasional deep moans and clattering of mandibles during a greeting ceremony and copulation. In northern India its breeding season is mainly from July to September. Thus, in mixed-species colonies as at Bharatpur and other places it has already completed its nesting when Painted Storks arrive to build nests in the same trees, from the end of August to early September. In fact, the latter use many of the abandoned nests of Asian Openbills as this species also builds platform nests – a collection of twigs with a central depression, usually lined with green plant material – as is often the case with other storks. Like the Painted Stork it, too, feeds the nestlings by regurgitating food onto the floor of the nest to be gobbled up by the chicks.

Ciconiini, comprising the 'typical storks', includes species belonging to genus *Ciconia*, many of which are true migrants, covering long distances across the globe, and having a mixed diet that includes insects, terrestrial arthropods, reptiles, amphibians and fish. The Asian species of interest in this tribe include the Black Stork *Ciconia nigra*, White Stork and Woolly-necked Stork

C. episcopus. Other storks in this tribe are Abdim's Stork *C. abdimii*, restricted to Africa; Storm's Stork *C. stormi*, restricted to certain parts of South-East Asia; Maguari Stork *C. maguari*, restricted to South America; and the Oriental Stork *C. boyciana* in East Asia.

Woolly-necked or **White-necked Stork** *Ciconia episcopus* **NT** (Figure 1.3).
The Woolly-necked Stork is slightly smaller than the Asian Openbill. Its iris is deep crimson or wine red in colour. This bird is glistening black overall with a black skullcap and a white, downy neck which gives it its name. The lower belly and undertail-coverts are white, standing out from the rest of the dark plumage. Feathers on the foreneck are iridescent with a coppery-purple tinge. These feathers are elongated and can be erected during displays. The tail is deeply forked and white, usually covered by the long black undertail-coverts.

FIGURE 1.3 Woolly-necked Stork. (N. K. Tiwary)

It has long red legs and a heavy blackish bill, though some specimens have largely dark red bills with only the basal one-third being black. Juvenile birds are duller versions of the adult, with a feathered forehead that is sometimes streaked black and white.

Sexes are similar, though males are thought to be larger. When the wings are opened during displays or for flight, a narrow band of very bright unfeathered skin is visible along the underside of the forearm. The Woolly-necked Stork is not a colonial nester and is usually found in pairs nesting individually in trees in agricultural landscapes. Small nestlings are pale grey with buffish down on the neck and a black crown. At fledging age, the immature bird is identical to the adult except for a feathered forehead, much less iridescence on feathers, and far longer and fluffier feathers on the neck. Newly fledged young have a prominent white mark in the centre of the forehead that can be used to distinguish that year's young. English common names for this species include the White-necked Stork, White-headed Stork, Bishop Stork and Parson-Bird. Its new English name of 'woolly neck' has been criticised by Indian ornithologists who maintain that wool is a term not frequently understood in India.

Black Stork *Ciconia nigra* **LC** (Figure 1.4).

The Black Stork has mainly black plumage with white underparts, long red legs and a long pointed, red bill. A widespread but uncommon species, it breeds in scattered locations across Europe (Portugal and Spain, and Central and Eastern Europe), and east across the Palearctic to the Pacific Ocean. It is a long-distance migrant, with European populations wintering in tropical sub-Saharan Africa and Asian populations in the Indian subcontinent.

Unlike the closely related White Stork, the Black Stork is a shy and wary species. It is seen singly or in pairs, usually in marshy areas, rivers or inland waters where it feeds on amphibians, small fish and insects, generally wading slowly in shallow water while stalking its prey. Breeding pairs usually build nests in large forest trees that can be seen from long distances, as well as on large boulders, or beneath overhanging ledges in mountainous areas. Despite its expansive range, it is nowhere abundant, and appears to be declining in some parts of its range, such as in India, China and parts of Western Europe, though increasing in others such as the on Iberian Peninsula. It bears some resemblance to Abdim's Stork, which can be distinguished by its much smaller build, predominantly green bill, legs and feet, and white rump and lower back.

The Black Stork has a wider range of calls than the White Stork, its main *chee leee* call sounding like a loud inhalation. It makes a hissing call as a

FIGURE 1.4 Black Stork. (N. K. Tiwary)

warning or threat. Displaying males produce a long series of wheezy raptor-like squealing calls, rising in volume and then falling. It rarely indulges in mutual bill-clattering when adults meet at the nest, or as part of their mating ritual or when angered. The young clatter their bills when aroused. The 'up–down' display – a common behaviour in storks in which one member in a pair moves its body up and down – is used for a number of interactions with other members (usually the mate) of the species.

Foraging for food takes place mostly in freshwater, though it may look for food on dry land also. It wades patiently and slowly in the shallows, often alone or in a small group if food is plentiful. It has been observed shading the water with its wings while hunting. In India, it often forages in mixed-species flocks with the White Stork, Woolly-necked Stork, Demoiselle Crane *Grus virgo* and

Bar-headed Goose *Anser indicus*. The Black Stork also follows large mammals such as deer and livestock, presumably to eat the invertebrates and small animals flushed by them.

White Stork *Ciconia ciconia* **LC** (Figure 1.5).
The White Stork measures on average 100–115 cm in body length. Its plumage is mainly white, with black on the wings. Adults have long red legs and long pointed, red bills. The White Stork is a long-distance migrant, wintering in Africa from tropical sub-Saharan Africa to as far south as South Africa, or on the Indian subcontinent. When migrating between Europe and Africa it avoids crossing the Mediterranean Sea and detours via the Levant in the east or the Strait of Gibraltar in the west, because the air thermals on which it depends for soaring do not form over water.

FIGURE 1.5 White Stork. (M. Mahendiran)

A carnivore, the White Stork eats a wide range of prey, including insects, fish, amphibians, reptiles, small mammals and small birds. It takes most of its food from the ground, among low vegetation and from shallow water. It is a monogamous breeder but does not pair for life. Both members of the pair build a large stick nest which may be used for several years. Each year the female can lay one clutch of usually four eggs, which hatch asynchronously 33–34 days after being laid. The parents take turns incubating and both feed the young. It has few natural predators but may harbour several types of parasites on its body.

The White Stork is gregarious; flocks of thousands have been recorded on migration routes and at wintering areas in Africa. They gather in groups of 40 or 50 during the breeding season. The smaller, dark-plumaged Abdim's Stork is often encountered with White Stork flocks in southern Africa.

Leptoptilini comprises the giant storks, basically in two forms: genera *Ephippiorhynchus* and *Jabiru* in one group and genus *Leptoptilos* in the other. The former includes the Black-necked Stork *Ephippiorhynchus asiaticus* of which there is an Asian as well as an Australian race. The Saddle-billed Stork *E. senegalensis* is resident in tropical Africa, while the Jabiru is found in the neotropical region. The second group in this tribe includes the adjutant storks (Lesser Adjutant *Leptoptilos javanicus* and Greater Adjutant *L. dubius*) which can cover long distances. Both species are found in Asia and are regarded as highly endangered. Their African counterpart is the Marabou *L. crumeniferus*.

The Shoebill *Balaeniceps rex*, also called Shoe-billed Stork or Whale-headed Stork, is normally placed in a monospecific family (Balaenicipitidae) and sometimes in its own order (Balaenicipitiformes) as its systematics are unclear.

Black-necked Stork *Ephippiorhynchus asiaticus* **NT** (Figure 1.6).
The Black-necked Stork is a tall, long necked species. It is resident across the Indian subcontinent and South-East Asia with a disjunct population in Australia. It lives in wetland habitats and near fields of crops such as rice and wheat where it forages for a wide range of animal prey. Adult birds of both sexes have a heavy bill and are patterned in white and iridescent black, but the sexes differ in the colour of the iris with females sporting yellow irises and males having red irises. This and the Saddle-billed Stork are the only stork species that show marked sexual dimorphism in iris colour. In Australia, it is sometimes called a Jabiru, although that name refers to a stork species found in the Americas. This is one of the few storks that is strongly territorial when feeding and breeding.

FIGURE 1.6 Black-necked Stork. The yellow iris indicates that it is a female bird. (N. K. Tiwary)

In India, the species is widespread in the west, central highlands and northern Gangetic plains, extending east into the Assam valley, but rare in peninsular India and Sri Lanka. This distinctive stork is an occasional straggler in southern and eastern Pakistan and is a confirmed breeding species in central lowland Nepal. It extends into South-East Asia, through New Guinea and into the northern half of Australia. The largest known breeding population occurs in the largely agricultural landscape of south-western Uttar Pradesh in India. During 1994–97, a total of six nests were recorded in Keoladeo Ghana National Park (Ishtiaq et al. 2004). This was their resident population size a few decades ago, but numbers now are likely to be far fewer.

Black-necked Storks forage in a variety of natural and artificial wetland habitats. They frequently use freshwater, natural wetland habitats such as lakes, ponds, marshes, flooded grasslands, oxbow lakes, swamps, rivers and water meadows. Small numbers are also seen in Indian coastal wetland habitats, including in mangrove creeks and marshes. In cultivated areas, they prefer natural wetlands to forage in, though flooded rice paddies are preferentially used during the monsoon, likely due to excessive flooding of lakes and ponds. Nests are usually in trees located in secluded parts of large marshes, or in cultivated fields as in India and lowland Nepal.

This large stork has a dance-like display. Both individuals stalk up to each other and, face to face, extend their wings and flutter the wing tips rapidly, advancing their heads until they meet. They then clatter their bills and walk away. The display lasts for a minute and may be repeated several times. In India, nest-building commences during the peak of the monsoon with most of the nests initiated from September to November and a few extending until January. They nest in large trees, sometimes isolated in sizeable marshes or in agricultural landscapes, on which they build a platform. In agricultural settings, human disturbance can cause nesting adults to abandon nests in some locations, but elsewhere they seem to nest successfully. The nest is large and made up of sticks and branches lined with rushes, water plants and sometimes with a mud plaster on the edges. Nests may be reused year after year.

Black-necked Storks are usually not intimidated and can be quite aggressive to other large waterbirds such as herons and cranes, and seem unafraid of raptors. This species is a carnivore and its diet includes some species of waterbirds and a range of aquatic vertebrates including fish (Maheswaran and Rahmani 2002), amphibians and reptiles, and invertebrates such as crabs and molluscs. They also prey on the eggs and hatchlings of turtles. In the Chambal River valley in Western India, they have been observed to locate nests of freshwater turtles and prey on their eggs.

Black-necked Storks sometimes soar in the heat of the day or rest on their hocks. When disturbed, they may stretch out their necks. Their drinking behaviour involves bending down with open bill and scooping up water with a forward motion followed by raising the bill to swallow the water. They sometimes carry water in their bill to chicks at the nest or even during nest-building or egg incubation stages, possibly to regurgitate it and bring the nest temperature down. Like other storks, they are generally mute except at nest where they make bill-clattering noises. The sounds produced are of a low pitch and resonate, ending with a short sigh. Black-necked Storks are largely non-social and are usually seen as single birds, pairs or in family groups.

Lesser Adjutant Stork *Leptoptilos javanicus* **VU** (Figure 1.7).
Like other members of its genus, this stork has a bare neck and head. It differs from them, however, in being more closely associated with wetland habitats, where it is solitary and is less likely to scavenge than the related Greater Adjutant. It is a widespread species found from India through South-East Asia to Java. The Lesser Adjutant is a fairly large stork, though smaller than its

FIGURE 1.7 Lesser Adjutant Stork. (Asad Rafi Rahmani)

cousin the Greater Adjutant and lacks the pouch in the neck. The skullcap is paler and the upper plumage is uniformly dark, appearing almost all black. The nearly naked head and neck have a few scattered hair-like feathers. The upper shank or tibia is grey rather than pink. During the breeding season, the face is reddish and the neck is orange. Males and females appear similar in plumage but males tend to be larger and heavier billed.

The Lesser Adjutant is often found on large rivers and lakes in well-wooded regions, on freshwater wetlands in agricultural areas, and coastal wetlands including mudflats and mangroves. It occurs in India, Nepal, Sri Lanka, Bangladesh, Myanmar, Thailand, Vietnam, Malaysia, Laos, Singapore, Indonesia and Cambodia, which has the largest population. In India the species is mainly distributed in the eastern states of Assam, West Bengal and Bihar. It may occur as a vagrant on the southern edge of Bhutan. These birds are extremely rare in southern India. In Sri Lanka they are found in lowland areas, largely within protected areas though they also use forested wetlands and crop fields. In Nepal, surveys in eastern districts suggest that they preferentially use forested patches with small wetlands, mostly avoiding crop fields. The Lesser Adjutant stalks around wetlands, feeding mainly on fish, frogs, reptiles, large invertebrates, rodents and small mammals; it occasionally consumes carrion. Location of prey appears to be entirely visual, with one observation being of these storks sitting on telegraphic poles apparently scanning a marsh for prey. Courtship behaviour of the Lesser Adjutant is identical to other species of the genus *Leptoptilos*.

Greater Adjutant *Leptoptilos dubius* **EN** (Figure 1.8).

Once found widely across Southern Asia, mainly in India but extending east to Borneo, the Greater Adjutant is now restricted to a much smaller range with only three breeding populations: two in India, with the largest colony in Assam and a smaller one around Bhagalpur, and the final one in Cambodia. They disperse widely after the breeding season. This large stork has a massive wedge-shaped bill, a bare head and a distinctive neck pouch. During the day it soars on thermals along with vultures with whom it shares the habit of scavenging. They feed mainly on carrion and offal; however, they are opportunistic and will

FIGURE 1.8 Greater Adjutant Stork nest. (Jaydev Mandal)

sometimes prey on vertebrates. The English name 'adjutant' is derived from their stiff 'military' gait when walking on the ground. Large numbers once lived in Asia, but they have declined (possibly due to improved sanitation) to the point of being endangered. The total population in 2008 was estimated at around a thousand individuals. In the nineteenth century they were especially common in the city of Calcutta, where they were referred to as the 'Calcutta adjutant' and included on the coat of arms for the city. Valued as scavengers, they were once depicted on the logo of the Calcutta Municipal Corporation. Known locally as *hargila* (derived from the Assamese words *har* meaning bone and *gila* meaning swallower, thus 'bone-swallower') and considered to be unclean birds, they were largely left undisturbed but sometimes hunted for the use of their meat in folk medicine.

The Greater Adjutant is omnivorous and, although mainly a scavenger, it preys on frogs and large insects and will also take birds, reptiles and rodents. It has been known to attack wild ducks within reach and swallow them whole. Greater Adjutants also catch fish, but their main diet is carrion.

Focusing on the Painted Stork

As an ornithological term, 'painted' which is affixed to the English name of the species, is not unique. Many bird names have this term as a prefix to their English names: Painted-snipe *Rostratula/Nycticryphes* spp. and Painted Spurfowl *Galloperdix lunulata* for instance. There is little in the ornithological literature that indicates how the term Painted Stork came to be used. Its earliest systematic description is by Thomas Pennant who, based on a type specimen collected from Sri Lanka, furnished the first account of the bird. According to the prevailing trends of ornithological nomenclature, the species was referred to as the 'White-headed Ibis' *Tantalus leucocephalus*. Pennant's description, reproduced below, includes no hint about how the bird could be visualised as it had been painted (Urfi 2011a). The illustration accompanying the account in Pennant's book *Indian Zoology* is a highly stylised figure.

> In size it is much superior to our largest curlews. The bill is yellow, very long, and thick at the base, and a little incurvated; the nostrils very narrow, and placed near the head; all the fore part of the head is covered with a bare yellow, and seems a continuance of the bill; and the eyes are, in a very singular manner, placed very near its base.
>
> The rest of the head, the neck, back, belly, and secondary feathers, are of a pure white; a transverse broad band of black crosses the breast; the

quil-feathers [sic], and coverts of the wings, are black; the coverts of the tail are very long, and of a fine pink color; they hang out and conceal the tail.

The legs and thighs are very long, and of a dull flesh color; the feet semi-palmated, or connected by webs as far as the first joint.

This bird was taken in the isle of Ceylon, and kept tame for some time at Colombo; it made a snapping noise with its bill like a stork; and, what was remarkable, its fine rosy feathers lost their color during the rainy season. (Pennant 1790: 11)

The Painted Stork has been referred to by other names in the past, such as Pelican Ibis (Jerdon 1864, Hume and Oates 1890). Clearly, the bird was thought to be a type of ibis initially and not a true stork. Its bald head and slightly decurved bill probably give the impression of its bearing a similarity to ibises. However, the realisation soon came that the bird in question was indeed a stork and the resemblance to ibises was only superficial. As Blanford (1898: 373) observes: 'This genus [*Pseudotantalus* and] *Tantalus*, which is an American form, chiefly distinguished by its naked neck, were long classed with the Ibises or in a family apart, but they are true storks.'

The scientific name of the bird has undergone a few changes over the centuries, from *Tantalus leucocephalus* to *Pseudotantalus leucocephalus*. From being placed under genus *Ibis*, it was finally placed under *Mycteria*, a genus which was erected by Linnaeus in 1758. The word is derived from the Greek word *mukter*, which means 'spout' or 'trunk', and the Latin *-ius*, which indicates 'resembling' or 'connecting'. The specific name derives from the Greek *luekok-phalos*, which simply means 'white headed' (*leucos*, white and *cephalus*, head). Thus, both the generic and specific names seem justified given that the Painted Stork has a large, heavy trunk-like bill and a white head (except for the bare skin of adult birds which turns orange in the breeding season). Its specific name was subsequently changed from *leucocephalus* to *leucocephala* to make the Latin genders match.

Comparative behavioural studies by Kahl (1970, 1971b, 1972a–d, 1973, 1974) revealed that Milky *M. cinerea*, Painted, Yellow-billed *M. ibis* and Wood Storks share extremely similar courtship patterns, particularly their up–down displays. Kahl suggested that all four of these species, then divided between the genera *Ibis* and *Mycteria*, should be combined into one genus, *Mycteria*.

That the storks belonging to the genus *Tantalus* (*Mycteria*) were all of a single group, with representatives in each of the continents of America, Africa and Asia, is something which was alluded to as early as 1864 by T. C. Jerdon. Of the Painted Stork, he noted that it 'is replaced in the Malay countries by *T. lacteus*, Temminck; and there are other species in Africa and America, *T. ibis*, and *T. loculator*' (Jerdon 1864).

Mycteria are all almost exclusively piscivorous and feed by tactolocation (see Box 6.2). From the viewpoint of genetic relatedness, Painted and Milky Storks exhibit 0.9% divergence in mitochondrial cytochrome *b*, indicating a recent split (Slikas 1997, 1998). In fact, the two species co-occur in some parts of South-East Asia and even interbreed in captive conditions (Urfi 2011a; Baveja et al. 2019).

Morphologically, the four species of *Mycteria* are distinct and there is little cause for confusion (Table 1.1). The Wood Stork is a comparatively smaller bird with a white body. Its head and neck are black and unfeathered with bony plates on the cap. Its legs are also black. The other three *Mycteria* congeners are all different from the Wood Stork but quite similar to each other. In all three the bill is waxy yellow and the legs pale red or flesh coloured. They have a patch of deeper yellow or orange skin extending from the base of the bill to the eyes and a little beyond. The size of the unfeathered patch, the 'bald head' so to speak, varies seasonally. In the Painted Stork it becomes very pronounced in the breeding season, while in the other two species it appears in sharper contrast against the bill.

In all of these three species (Figure 1.9) the body plumage is white with black in varying degrees, especially on the primaries. The Milky Stork is wholly white with a rim of black feathers on the wings. The Yellow-billed and Painted Storks have more patches of black, as well as pink feathers, which are lacking in the Milky Stork. In the Yellow-billed Stork's case, the pink is more pronounced on the back and shoulders; but in the Painted Stork, pink is restricted to the tail region and appears only in the breeding season. Morphologically, these two species also differ in another important respect which is that the Painted Stork has a broad pectoral band that is absent in the Yellow-billed Stork.

As will be clear from Table 1.1, the geographical distribution of each of the four species is different. The Milky Stork is currently considered endangered, with about 1,500 individuals left in the wild (BirdLife International 2023b) and its populations rapidly declining (Li et al. 2006) due to widespread habitat destruction and hunting (Verheugt 1987; Iqbal and Ridwan 2008; Iqbal et al. 2009).

TABLE 1.1 *Mycteria*: A comparative study (Figures 1.9)

Species	Conservation status	Appearance	Distribution	Nesting ecology
Wood Stork *Mycteria americana*	LC	Black head and bill. Bill decurved like all *Mycteria*. Overall white plumage.	Some south-eastern states of the USA; Mexico, through Central and South America to northern Argentina.	Dry-season nester. Breeding initiated by drop in water level combined with increased concentration of fish (former likely triggering latter) anytime between November and August.
Yellow-billed Stork *Mycteria ibis*	LC	Bill is deep yellow, slightly decurved at end; rounder cross-section than storks outside *Mycteria*. Feathers extend onto the head and neck just behind eyes; face and forehead covered by deep red skin. Sexes similar in appearance.	Restricted to the African continent, south of Sahara, and Madagascar; occasionally in Morocco, Tunisia and Egypt.	Breeding is seasonal; appears to be stimulated by the peak of long, heavy rainfall and resultant flooding of shallow marshes, usually near Lake Victoria. This flooding is linked to an increase in fish prey availability.
Painted Stork *Mycteria leucocephala*	NT	Head of the adult is bare and orange or reddish in colour. Long tertials tipped bright pink which at rest extend over back and rump. Distinctive black breast band with white scaly markings. The band continues into the underwing-coverts; white tips of black coverts give it appearance of white stripes running across the underwing lining. Rest of body whitish in adults. Primaries and secondaries black with greenish gloss. Legs yellowish to red but often appear white due to habit of defecating on their legs, especially when at rest. Short tail black with green gloss.	Stronghold in Indian subcontinent; range spreads eastwards to Indo-China.	Wet season nester after monsoon rains during which fish breed.

(continued)

TABLE 1.1 *Mycteria*: A comparative study (Figures 1.9) (continued)

Species	Conservation status	Appearance	Distribution	Nesting ecology
Milky Stork *Mycteria cinerea*	EN	Slightly smaller than the closely related Painted Stork. Adult plumage is completely white except for black flight feathers of wings and tail, which also have a greenish gloss. Extensive white portion of plumage completely suffused with pale, creamy yellow during breeding season, hence 'milky'. Wing coverts and back feathers are paler and have off-white terminal band. Bare facial skin greyish or dark maroon with irregular, black blotches. During breeding, bare facial skin is deep wine red; black markings on the lores by bill base and gular region; ring of brighter red skin around eye. Breeding birds also show narrow pinkish band of bare skin along underside of wing. Downcurved bill dull pinkish-yellow, sometimes tipped white.	Occurs in south Vietnam, Cambodia, peninsular Malaysia, and Indonesian islands of Sumatra, Java and Sulawesi. In Cambodia, Malaysia, and Vietnam it co-occurs with Painted Stork but latter absent from Indonesia.	Typically breeds in dry season after the rains, April to November. Onset of breeding varies throughout species' range but probably coincides with maximum fish stocks and density following fish reproduction in the rainy season.

FIGURE 1.9 Distinguishing features of the four species of *Mycteria*: (top left) Wood Stork; (top right) Yellow-billed Stork; (bottom left) Painted Stork; and (bottom right) Milky Stork. (Wood Stork © Mike's Birds, Wikimedia Commons, Creative Commons Attribution-Share Alike 2.0; Yellow-billed Stork © Derek Keats, Wikimedia Commons, Creative Commons Attribution 2.0; Painted Stork by Paritosh Ahmed; Milky Stork by Debjyoti Ghosh)

Hybridisation between Painted and Milky Stork in South-East Asia

Hybridisation leading towards extinction of endangered species is one of the major threats to confront in conservation biology (Todesco et al. 2016). Usually, the genomic composition of endangered populations gets compromised by the infiltration of alien alleles into the native gene pool leading to hybrid swarms (Allendorf et al. 2001), species collapse (Kleindorfer et al. 2014) and eventual extinction (Rhymer and Simberloff 1996; Wolf et al. 2001). Through the process of introgression, endangered species with very small populations can eventually get absorbed into the genome of the more widespread species (Baveja et al. 2019). There is evidence that hybridisation between the Milky and Painted Storks is happening at some places in South-East Asia under captive conditions. Even though the ranges of the two species have historically overlapped, reports of hybridisation in the wild have only started appearing since the recent drastic decline of the population of Milky Storks (Baveja et al. 2019). These hybridisation events are presumably

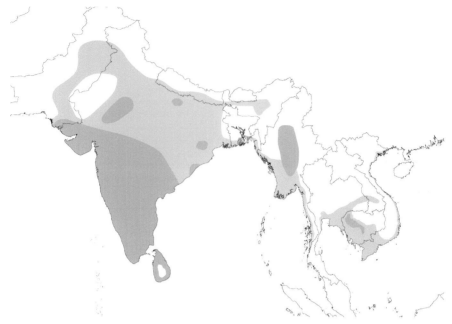

FIGURE 1.10 Painted Stork in South Asia (India, Pakistan, Sri Lanka, Myanmar, Bangladesh and other countries) and South-East Asia (Malaysia, Indonesia, etc). Light blue areas denote non-breeding and the darker blue areas denote year-round breeding. (Reproduced with permission, Cornell Lab of Ornithology)

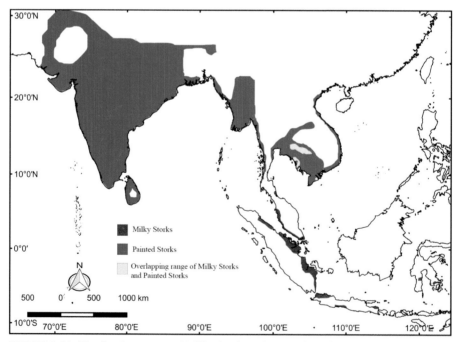

FIGURE 1.11 Distribution range of Milky Stork and Painted Stork in South and South-East Asia. Courtesy Dr Pratibha Baveja (for details, see Baveja et al. 2019).

due to limited mate choice in mixed nesting colonies (Li et al. 2006; Hancock et al. 1992), hence putting the Milky Stork in further peril. Figure 1.12, below, illustrates a hybrid between a Painted Stork and a Milky Stork. The plumage is clearly intermediate.

Distribution of the Painted Stork

India

Generally speaking, food availability is the single most important factor governing the ecology, distribution and biology of storks. Storks are good examples of birds that are reproductively limited by food (Lack 1968), and it is in this context that we must view the distribution of the Painted Stork across its range. For those birds that are specialised foragers, certain conditions with respect to food availability, abundance and type must be met if they are to breed successfully in any given region. Thus, the presence of an individual in any locality does not necessarily imply readiness for breeding, which is heavily dependent upon food cycles and other factors. In many India-specific works such as Ali and Ripley (1987), Kazmierczak (2000) and Grimmett et al. (1998), as well as in texts covering the biology of storks in detail such as del Hoyo et al. (1992) and Hancock et al. (1992), the distribution of the Painted Stork

FIGURE 1.12 A hybrid between Painted and Milky Storks. (Debjyoti Ghosh)

is shown as being spread across a very wide area, extending from Pakistan in the west to areas in South-East Asia. Grimmett et al. (1998) and Rasmussen and Anderton (2005) have provided distribution maps in their works, showing a more restricted distribution for the species. We will consider the distribution of Painted Stork breeding centres in detail in Chapter 3, but here we concern ourselves with the overall distribution of the species throughout its range (Figure 1.10).

From the western limit of its range in Pakistan, the Painted Stork is mainly recorded along the River Indus and in the major lakes in Sind. It is spotted in Punjab along the larger river channels on the plains. Some recent sightings of Painted Storks in Pakistan have been at Rawal Lake, Thatta, Badin and the Balloki headworks on the River Ravi. Most authors from Pakistan are ambiguous with regard to nesting and no specific site seems to have been recorded in recent times. While Roberts (1991) mentions that the bird breeds on the Indus delta, he does not furnish any site details. Among some very old nesting records from areas that were in former times within British India but now lie in Pakistan is an account of a colony from East Narra in Sind (Urfi 2011a).

East of the River Indus, close to the India–Pakistan border, there appears to be a zone from where no Painted Stork sightings have been reported (Rasmussen and Anderton 2005), probably due to the presence of the Thar Desert. It is only when one reaches southern and eastern parts of Rajasthan that Painted Storks become abundant. However, recent reports suggest that the local ecology and landscape are changing drastically due to the construction of the Indira Gandhi Irrigation Canal (completed 2005–10) and the number of waterfowl has increased. It is also likely that since so few birdwatchers frequent these areas, due to the desert conditions, some under-reporting is to be expected.

From the northern part of the Indian subcontinent, in the Indian Punjab (as well as parts of Pakistani Punjab) the Painted Stork is well represented. In fact, 600 were reported for the Asian Waterbird Census (AWC) of 1990 at a site known as Lehalan. In Nepal the Painted Stork is a rare summer visitor and resident below 250 m elevation. As far as China is concerned, the species was previously reported as a common summer visitor in the south, probably breeding, but now it is rare and possibly extinct (BirdLife International 2023a).

In India, Painted Storks occur across the length and breadth of the country with a few exceptions. For instance, in Kerala, in the extreme south, and in northeastern India it is a rare visitor. In Assam, Painted Storks are rare and sighted only sporadically, mostly in the Bhramaputra River valley. In Bangladesh the bird is not common, while in Myanmar it was formerly resident in the central region and a visitor throughout. According to Smythies (1940), it was 'a very common bird on the coastal mudflats, seen both solitarily as well as in flocks'; it also frequents *jheels* (marshes and wetlands) and flooded paddy fields. The current status of the Painted Stork in Myanmar is largely unknown, but it is perceived as a very rare species there (BirdLife International 2023a). In the 1992 AWC only four individuals were recorded, suggesting that its occurrence there was highly sporadic at that time. There is no record of Painted Stork from the Andaman and Nicobar Islands in the Bay of Bengal. Sri Lanka is, of course, the second most important country for Painted Storks, and here they are locally abundant, particularly in the dry zone (Henry 1971; del Hoyo et al. 1992; Grimmett et al. 1998; Kaluthota et al. 2005).

South-East Asia

The Painted Stork has been reported from several countries in Indo-China (Figures 1.10 and 1.11). From a global perspective, it would seem as if its distribution is discontinuous, but this is likely to be only because in large parts of this region the bird has been persecuted or its habitat has come under threat and hence it has become absent from large areas. Thus, this may not actually be a case of true discontinuous distribution in a biogeographical sense (Ripley and Beehler 1990).

In Laos, the Painted Stork was previously widespread but is now rare. In Vietnam, also, it was formerly a widespread resident but has become a rare nonbreeding visitor (Grimmett et al. 1998; Robson 2000). In Malaysia it is currently regarded as a vagrant (Wells 1999; Robson 2000) and is to be found in freshwater marshes, lakes and reservoirs, flooded fields, rice paddies, swamp forest, riverbanks, and on intertidal mudflats and salt pans. According to Delacour (1947), it was previously found in the northern Malay states. It is only in Cambodia that the Painted Stork seems to be doing well. Here it is a local resident, with a minimum of several hundred pairs breeding at Tonle Sap, which is situated within the floodplain of the Mekong River and is the largest natural freshwater lake in South-East Asia. The Tonle Sap Great Lake colony is probably the largest in South-East Asia (Campbell et al. 2006) and an estimated 20% of the regional Painted Stork population exists here. The site is also important for the only freshwater colony of Milky Stork in the world. A small colony of Painted Storks has also been reported at Ang Trapeang Thmor Sarus Crane Reserve, where hybridisation with Milky Storks is also suspected.

Finally, in Thailand the Painted Stork was previously regarded as a common breeder in the south but now it is on the verge of extinction, and only recorded sporadically in small numbers (Wells 1999; Robson 2000).

Representations of the species

Due to my lifelong engagement with Painted Storks, I have always been on the lookout for representations of this species (and birds in general) in public spaces. Goods vehicles caught my attention because they often have avian designs painted on their sides, associated with messages, idioms or poetry reflecting on the lonely and difficult life of a truck driver or the forces of good and evil. The makers of truck carriages seem to choose images of bird species found in the open countryside and not those of jungles, for depiction on the sides of the trucks. In fact, the commonest images are from only two groups: pheasants and raptors, which in India are considered as glamorous birds. Representations of the Indian Peafowl *Pavo cristatus*, the national bird, are ubiquitous. Raptor images are more generalised and can be a kite, eagle or hawk. Truck drivers are likely to encounter wetlands as they move around in the country, where large fish-eating birds – waders such as herons, egrets and storks – are commonly seen. However, the closest to a Painted Stork that I have seen is one depiction of an egret on the body of a goods-ferrying truck (Figure 1.13).

Having said this, pleasant surprises concerning pictorial representations of the Painted Stork do turn up (Figure 1.14). At Kokkare Bellur village in South India, which will be discussed in detail at several points in this book, locally nesting Painted Storks and Spot-billed Pelicans *Pelecanus philippensis* enjoy enormous goodwill among the local people and local buildings have depictions of these birds on their walls (Figure 1.15). Painted Storks are also represented in handicrafts and paintings and on postage stamps. In India, three such stamps have appeared, one of them being the se-tenant postage stamp issued by the Indian Postal Department in 1996 to mark the centenary of the birth of Dr Salim Ali, the country's most prominent ornithologist. The scene depicted in the stamp is that of Keoladeo Ghana National Park in Bharatpur, which has extensive colonies of the Painted Stork. Dr Salim Ali was instrumental in getting the park declared as a protected area. A special cover depicting Painted Storks issued by the Indian government's Post and Telegraphs Department is shown in Figure 1.16. The species has also appeared on postage stamps of Sri Lanka, Vietnam, Thailand, Cambodia and several other countries.

FIGURE 1.13 Bird images on trucks in India are usually of species commonly encountered in the open countryside and not those of closed habitats such as jungles and woodlands. Depictions of fish-eating waders like egrets are occasionally seen. The photograph shows an image of an egret – a relative of the storks – painted on the side of a goods-ferrying truck. (A. J. Urfi)

FIGURE 1.14 An image of a Painted Stork on the railway station of Bharatpur, indicating that this town in Rajasthan is identified with the world-famous bird sanctuary and its waterbirds. (A. J. Urfi)

FIGURE 1.15 The nesting birds at Kokkare Bellur village, notably Spot-billed Pelicans and Painted Storks, enjoy enormous goodwill among the local people. A local building has a drawing of these birds on its walls. (Bharat Bhushan Sharma)

FIGURE 1.16 First-day cover with images of Painted Stork issued by India's Post and Telegraph department. (Rahul Rohitashwa, personal collection)

BOX 1.1 Some important stork researchers

Storks are good models for studying some questions in biology and ecology but there are certain limitations. The list of scientists who have worked on storks is not a long one, but it includes many names from the Americas. This is because the Wood Stork, a denizen of the southern parts of the United States and some South American countries, has received considerable attention from ornithologists with respect to its ecology, genetics, behaviour and so on. Research papers on different aspects of the biology of the Wood Stork have resulted from the work of several groups in the southern United States and also Mexico and Brazil.

While it is not my intention to provide a comprehensive list of all stork workers here, I will mention a few whose research is regarded as foundational and with whom I have had some degree of personal association. The primary person who made a seminal contribution to the study of storks was Philip Kahl (1934–2012). Phil started his graduate studies in the 1960s working in the Corkscrew Swamp Sanctuary which has the largest Wood Stork colony in the United States (Hill 2014). His experiments with captive-reared Wood Storks led to significant papers on bioenergetics. But his most famous work was the discovery of the 'bill-snap reflex' mechanism and his paper on this subject, published in the journal *Nature* (Kahl and Peacock 1963), was also reported in the science section of *Time* magazine.

Another of Kahl's landmark papers described 'urohidrosis', a thermoregulatory mechanism in some species of birds to prevent hyperthermia by excreting urine on their own legs. This behavioural adaptation helps birds to get rid of excess body heat as the water in the excreted urine evaporates.

After completing his PhD at the University of Georgia in 1963, Kahl went on to study storks in Africa and Asia, and from 1959 to 1969 he studied the breeding behaviour of all 11 of the then recognised species of storks in the world. His works led to advancements in stork taxonomy, with particular reference to the genus *Mycteria*.

Another ornithologist who deserves a mention in the context of stork research is James Hancock OBE (1921–2004). Having served in British India he belonged to that vanished tribe of Englishmen who had a connection with India while it was still a British colony (Kushlan 2006). Today he is remembered for producing, along with James Kushlan and Philip Kahl, the well-regarded book, *Storks, Ibises and Spoonbills of the World*.

Illustration of a Painted Stork from *Stork, Ibis and Spoonbills of the World* by Hancock et al. (Artists: Alan Harris and David Quinn). This painting is indeed a very realistic portrayal of the Indian agricultural landscape – a farmer ploughing his fields with bullocks pulling the plough, a line of trees in the far distance, green fields all around and in the sky a pair of Painted Storks flying. This is the sort of scenery, maybe of the flat Indo-Gangetic plains, one might see through the window of a train. (A. J. Urfi, personal collection)

Chapter 2

Avian Coloniality

Nesting behaviour patterns among birds vary a great deal. Simply speaking, most birds, notably the passerines, defend a territory and build a single nest in a tree or some other suitable location. This nest may be quite far away from a nest belonging to another bird of the same species. Such birds are highly territorial and by employing a variety of methods, including song and dance, they advertise an 'all-purpose territory', a patch in their habitat which has all the necessary resources – such as food, water, nesting spaces – needed for survival and reproduction (Figure 2.1). On the other hand, some birds build their nests quite close together in a colony in the same tree. So, the question

FIGURE 2.1 At the commencement of their breeding season, most birds defend an all-purpose territory using a variety of methods, including song, dance or display of plumage. The Oriental Magpie Robin *Copsychus saularis*, a common garden bird in India, is highly territorial. (N. K. Tiwary)

is, why do some bird species nest singly while others are colonial? This is a question that has vexed ornithologists for a long time, particularly in context of colonial waterbirds (Perrins and Birkhead 1983; Urfi 2003b).

Many types of waterbirds, particularly species of heron, egret, stork, ibis, spoonbill, cormorant and pelican, build nests on trees in the middle of a marsh; these sites are known as 'heronries'. At such sites one can see quite clearly that the birds have congregated on just a few trees while ignoring others that seem similar (Figure 2.2). The selected trees could be good in terms of their suitability as a nesting substrate, but the question remains: with dozens of other trees in vicinity, why place nests in tight groups? British naturalists, who laid the foundations of ornithology in India, could barely conceal their bewilderment at the sight of heronries. Betham (1904), an ornithologist in British India expressed this clearly when he wrote, 'It is curious why waterbirds should breed in colonies; where trees are scarce, it might be understood, but where there are plenty standing in water, why they should pack together is hard to understand. It must be miserably uncomfortable to have no elbow room.'

It was the British ornithologist David Lack who, in his book *Ecological Adaptations for Breeding in Birds* (Lack 1968), was the first to consider questions such as these in an evolutionary context (Perrins and Birkhead 1983). He suggested that the major selective factors favouring colonial nesting were successful avoidance of predators and more efficient exploitation of food resources. Lack estimated that approximately 13–14% of all birds are colonial, although this estimate is most probably conservative.

FIGURE 2.2 In this picture it is clear that Painted Storks nests are grouped tightly in only some clumps of trees while other trees seem to have been ignored. Ecologists have asked the question why this should be so? (A. J. Urfi)

The tendency to breed in colonies seems to be more common in waterbirds. However, as a taxonomic group not all waterbirds (including some storks) are colonial and so for finding answers to the evolution of coloniality we will have to look elsewhere, possibly to the ecology of particular species and the way the resources that birds utilise are distributed in their environment. But as a survival strategy, coloniality must have certain benefits for it to have evolved through natural selection. While there have to be costs associated with nesting in the company of other birds (both of the same species or others in mixed colonies), there must be some advantages, too. The benefits must surely outweigh the costs.

Let us have another look at a typical Painted Stork colony. Figure 2.3 shows one such colony with nests on clumps of acacia trees growing on an island in the middle of a large village pond. The location on an island obviously confers an advantage to the nesting birds because it is harder for predators, including humans, to get to it and devour nest contents; this must have been a very important selective force in the evolution of coloniality. This line of reasoning perhaps explains why no Painted Stork nests are to be seen on trees in the far background in Figure 2.3. Since they are not surrounded by water, it may be easier for predators to reach them and climb up those trees to get to the eggs or helpless young ones in the nests.

While this partly explains colonial nesting, what about other dangers, such as predation from the air? With a concentration of so many nests, predatory

FIGURE 2.3 Painted Stork colony on an island. It would seem that the water barrier around the nesting colony provides some degree of protection from land predators. (N. K. Tiwary)

birds such as raptors are likely to have a feast. So, perhaps, it is not so simple? In fact the reasons for the evolution of coloniality are many and complex. To ward off aerial predators, birds have to be constantly on their guard, and with so many nests, each with two or atleast one watchful parent, at any given point in time there are likely to be a few pairs of eyes on the lookout to give advance warning of the approach of any raptor (Brown and Brown 2001; Urfi 2003b).

Breeding ecology of birds

Food availability is believed to be one of the most critical resources for repro-duction. The breeding patterns of birds are intimately tied to the distribution and abundance of food resources in their environment. We can compare and contrast this situation with mammals, higher vertebrates that are also warm-blooded creatures like birds. In mammals, milk production requires only a sufficient amount of food supply for the adults, but in the case of birds, which feed their young on food more or less as it is – in other words, without any processing or pre-digestion (with some exceptions) – the food supply has to be both abundant and constant. As peaks in food abundance can be very short, the breeding seasons of birds generally tend to be highly localised in time and space, compared with those of mammals.

A well-known theory in ecology known as the 'food availability–breeding time' hypothesis suggests, rather logically, that most birds breed at that time of the year when plenty of food is available for their chicks (Perrins and Birkhead 1983). This theory was developed by David Lack and several of his students at the Edward Grey Institute of Field Ornithology, Oxford University (see Chapter 6 for more on this). Seen in the light of evolutionary ecology, which attempts to explain survival strategies in terms of direct or indirect enhancement of their reproductive success (Darwinian fitness), this theory makes sense and has fairly wide applicability.

We can begin by attempting to have a general understanding about food resource distribution in the environment and how it impacts survival strategies of different species of birds, which in turn affects their nesting habits. The primary thing is that food is not evenly distributed in space and time. Its avail-ability and distribution are linked to a host of factors. For instance, in India, for many frugivorous birds such as barbets and parakeets, the time when trees come into flower and fruit may be synchronised with climatological patterns. Indeed, over much of India it is the monsoon which exerts a key influence over the food cycles of birds (Padmanabhan and Yom-Tov 2000). In a beautiful essay, published in a local newspaper, on the nesting habits of Indian birds entitled 'Stopping by the woods on a Sunday morning', Salim Ali, the famous ornithologist, wrote: 'The monsoon is the breeding season par excellence of insectivorous birds, and also of the numerous others who, though when adult, subsist principally on grain, yet require soft food in the nature of juicy grubs and caterpillars to nourish their young in the nest' (reproduced in Urfi 2008). Considering that the major nutritional requirement for growing birds is protein, which is best available from live animals, this statement makes a lot of sense. Much like for terrestrial habitats, in wetlands, too, it is the monsoon that plays a major role in triggering food cycles.

In order to understand how birds have evolved different types of nesting patterns, ornithologists have used a variety of approaches, the most helpful of which have been to build simple models based on optimality principles, which we will discuss below. To make our task easier let us restrict ourselves to two contrasting types of nesting patterns found in the world of birds, namely solitary nesting and colonial nesting. A number of birds such as barbets, owls, hornbills, sparrows, mynahs and bulbuls build nests in isolation, while group-nesting or colonial-nesting is most prominently exemplified by some, although certainly not all, waterbirds.

The economics of solitary and colonial nesting

One simple though old-fashioned way to start a discussion about different food exploitation patterns is the Economic Defendability Model proposed by Jerram Brown in 1969 (Perrins and Birkhead 1983; Urfi 2003b). Our starting premise is that resources (primarily food) are not evenly distributed in space and time. In fact, only three situations are possible, as shown in Figure 2.4. The first situation, in which resources are randomly distributed (A), is not very realistic because resources like food, nesting material or availability of mates are never randomly distributed. Typically, these resources are tied to physico-geographic features of the landscape and are distributed along certain well-defined patterns in the natural world, across different spatial scales. Consider a forest in which there are a number of trees attracting insects, birds and other fauna due to their fruits, flowers (or the insects attracted to them) and also the substrate which the trees themselves offer in terms of a habitat (say, for nesting). The distribution of the trees will not be random because they themselves are distributed along some well-defined gradients of soil, moisture and other factors. Therefore, the second situation (B) in which resources are regularly distributed is slightly more realistic, but not completely so. The third situation (C) is one in which resources are clumped in their distribution.

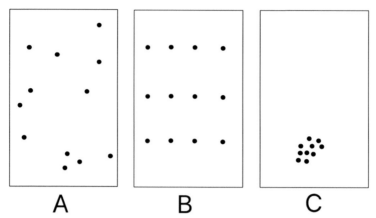

FIGURE 2.4 Graphical description showing the three main types of resource distribution patterns found in the natural world. (Redrawn from Urfi 2003b)

Imagine shoals of fish in an ocean. Such resources are neither distributed randomly nor regularly. There are vast empty spaces where there are no fish and then suddenly a shoal is encountered. From the viewpoint of a predator such as a seabird, vast distances have to be covered to track and locate these highly clumped food resources. But once located there is no real scarcity because of the local super-abundance of fish.

Besides being localised in space, food or other resources can also be clumped in time. For instance, imagine a forest bird that depends upon the nectar of flowers and, therefore, the moment when the tree comes into its flowering phase (Figure 2.5). Since this may happen only at one particular time of the year, the local resource availability, from the point of view of the bird, will be crucial for its survival. Given these basic patterns of resource distribution, we can now ask the question: what sort of nesting patterns should evolve in different environments? In Figure 2.6, again consider the black dots as resources and assume that in any area a pair of birds require 6 units (dots) of resources to survive and reproduce. In situation A, the resources are densely and fairly uniformly distributed. Consequently, five pairs of birds can survive in this area, each with a small territory (outlined with circles) that can be defended with ease. The bird in question expends considerable energy in the defence of this territory, fighting with rival males to obtain exclusive rights over it.

In fact, in situation B, with resources distributed sparsely and somewhat spaced out, territories are large and energetically costly to defend. So, another way to look at this situation would be to ask the question: what should be the optimal territory size for a bird? This is best described by the Economic Defendability model (Figure 2.7), which hypothesises that as territory size

FIGURE 2.5 Resources are clumped in space and time. A tree's resources such as nectar in flowers may only be available for a specific time and limited duration each year. (A. J. Urfi)

increases, both the costs and the benefits increase, but in different ways. The costs may be in terms of energy used to defend the territory from conspecifics, and the benefits may include the reserves of food or nesting material that can be obtained with ease within a defined area. The model assumes that the costs increase more or less linearly but the benefits increase exponentially at first and then level off. Clearly, larger territories have more resources but are difficult to defend and beyond a point, no matter how large the size of territory, the

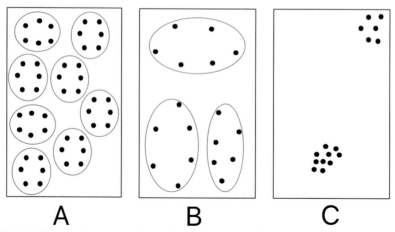

FIGURE 2.6 Territory formation in relation to resource distribution. **A.** An abundance of resources results in several individuals forming small, compact territories; **B.** Due to the scarcity of resources fewer territories are possible, leading to significant demographic changes. As shown here three birds will be able to survive and each with a comparatively larger territory perimeter to defend. The effort (in terms of energy and time) needed for this defence, however, may not always be justified by the quantity of resources defended; **C.** Resource distribution is highly clumped. If a pair of birds successfully defended this superabundance, much of the resource would go unutilised. Moreover, it would be pointless to defend a resource which is located far away from the nest, such as in the middle of a large water body. (Redrawn from Urfi 2003b)

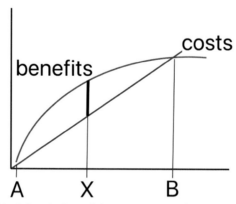

FIGURE 2.7 A graphical description of the Economic Defendability Model. As territory size increases, both costs and benefits increase, but in different ways. The costs, such as energy needed for the defence of territory increase in a linear manner with increasing territory size. However, the benefits (e.g., the amount of food) increase very rapidly at first but level off over a certain territory size. Beyond a point it doesn't matter how abundant resources are because the organism can only consume as much as is required. According to the model, territories of a size between A and B are worth defending, and the optimum territory size is at X where the benefit:cost ratio is maximum. (Redrawn from Urfi 2003b)

resources that will be used by the owner can only be protected to a certain capacity. It is pointless to defend a very large territory.

So, different patterns of nesting behaviour are likely to evolve according to the type of resource distribution. For a barbet inhabiting a forest where resources are more or less regularly distributed, a territorial system seems logical. It is not uncommon to witness the *tut-tut-tut* calls of Coppersmith Barbet *Psilopogon haemacephalus* on hot summer afternoons in most parts of India. As you walk along you hear the male's familiar metallic call, sounding like a hammer in a blacksmith's workshop, advertising his territory (Figure 2.8). But as soon as one individual has started to advertise, another begins calling, almost as if to say, 'Hey, I know you have a territory but I have one, too. Don't you dare encroach upon mine.' Very soon the forest is ringing with the multitudinous voices of barbets.

In contrast to this situation, if the food is highly clumped in space and time, territoriality can break down. Imagine a flock of gulls prospecting for fish in the sea. Once the school is located the gulls get busy harvesting this resource. Since food is abundant there is no point in defending this resource, which is anyway impossible to defend due to being in the middle of the sea. Moreover, since the food is located at a place where nests cannot be built (all birds are

FIGURE 2.8 Coppersmith Barbet. (N. K. Tiwary)

terrestrial animals and require land for nesting purposes), the nesting and feeding areas are far apart. In this situation the territorial mode of nesting would not work. Instead birds nest in colonies and more or less collectively go to the foraging areas by watching where others go and following suit. In this way, breeding colonies also act as information centres.

Theories of avian coloniality

The theories for the evolution of avian coloniality are several and the last word on this subject has not yet been said (Siegel-Causey and Kharitonov 1990; Brown and Brown 2001). Instead of providing a scholarly review of all the ideas and the current status on coloniality, here we will just highlight a few major themes. Like other forms of nesting, coloniality has its advantages and disadvantages for the individuals who choose to adopt this strategy, and so a common approach has been to list out all the costs and benefits of nesting together (Figure 2.9). In this approach, many variables can enter the argument simultaneously as a 'cost' and as a 'benefit'. For example, energetics (**e**, energetic effect) can enter the argument as a benefit because colonial nesting is certainly believed to enhance food finding and foraging efficiency. It has been postulated that some transfer

FIGURE 2.9 Breeding in close proximity to others, individuals belonging to the same species or different species, presents both challenges and opportunities. (Rajneesh Dwevedi)

of information about food resources takes place in a colony, and if one bird locates a food resource, others can then follow it and exploit the same resource. In fact, some ornithologists have gone so far as to suggest that the plumage patterns of certain waterbirds may have evolved to facilitate communication regarding food location. However, the energetic aspects of food finding can also enter the argument as a cost because competition, including stealing of food, occurs both at the food patch and at the nesting site.

Another pressure on birds nesting in colonies is danger from predation (**p**, vulnerability to predation) and this too can enter the argument both as a cost and a benefit. Many evolutionary biologists have discussed how group living allows group defence against predators, and how the chances of an individual being predated upon decrease when it lives in flocks or groups (Hamilton 1971); however, its disadvantage is that large groups of nesting birds are more visible to predators (Frederick and Meyer 2008; Frederick and Spalding 1994).

One compelling aspect of nesting in colonies is that these may be the only places where individuals can find mates (**m**, mating opportunities), because in the non-breeding season, each individual bird, whether male or female, leads a solitary existence. In fact, the nesting colony may be the only place where an individual could find a mate because these birds do not use acoustic or scent signals to attract individuals of the opposite sex. Having said this, at a nesting colony there is likely to be a large number of potential nesters and there will be stiff competition among them (the males) for suitable nesting sites and mates. Not all potential breeders will be successful. However, while the potential for finding a mate is a benefit there can be costs attached. For instance, at the colony site there may be intense competition for nesting sites and increased potential for theft of nest material (**i**, interference perpetrated by neighbours). There may also be more chance of extra-pair copulations and so each individual may spend more time and effort in guarding against it (Westneat et al. 1990).

Wittenburger and Hunt (1985), in their review on the evolution of coloniality, consider the net effect of coloniality on each significant variable. They have argued that coloniality should evolve when:

$N_e + N_p + N_m + N_i > 0$, where

- N_e is the benefit from enhanced foraging efficiency minus the cost of increased competition (net *energetic effect*),
- N_p is the benefit from enhanced defence minus the cost of increased attraction of predation (net effect on *vulnerability to predation*),
- N_m is the benefit of increased access to mates minus the cost of increased competition for mates (net effect on *mating opportunities*),

- N_i is the benefit from increased opportunities to exploit or disrupt neighbours (for example, stealing nest material or killing chicks) minus the increased degrees of *interference* perpetrated by neighbours.

While the cost–benefit analysis offers one explanation of the evolution of coloniality, there are other views too in ornithological and ecological literature. At the other extreme is the theory of 'commodity selection', advanced by French evolutionary biologists Danchin and Wagner (1997) and Wagner et al. (2000). These authors seem to argue that there is no such thing as coloniality and the reason why waterbirds and some other species nest in groups is that the resources they depend upon are themselves clumped. So what looks like colonial nesting is actually an artefact of this. For instance, for some species suitable nesting sites may be scarce. Colony sites may be the only places to find a mate. The theory of commodity selection has been strongly contested and debated (see Tella et al. 1998).

Meanwhile, other interesting ideas have also appeared. Jovani and Tella (2007), for instance, present a unique perspective in their paper, 'Fractal bird nest distribution produces scale-free colony sizes', by analysing the spatial distribution of White Stork nesting patterns across different spatial scales across continental Europe. In their review of avian coloniality, Brown and Brown (2001) attempt to define what constitutes a colony from the perspective of behaviour of birds. According to these authors, 'unlike with territoriality, which can be explained in terms of economic defendability or cooperative breeding, which can be understood in terms of inclusive fitness theory and habitat limitation, work on coloniality (and communal roosting) has not uncovered general rules that can easily explain either its evolution or maintenance.'

Some terms explained

Active coloniality is when individuals choose to nest in group; that is, in close proximity with each other, whether in the same tree (substrate) or on the ground. However, not all cases of group nesting are instances of active coloniality. In some cases, a critical factor, say the nesting substrate may be limited or very scantily distributed. In such a situation, while colonies would form, they would be due to limited availability of nesting sites, not because of any benefit conferred by colonial nesting. Such a situation might be illustrated by seabirds that nest on islands in the sea (Coulson and Dixon 1979; Kharitonov and Siegel-Causey 1988). Since there may be only a few suitable islands (others may not be suitable due to the presence of predators or human disturbance), birds may be forced to nest together because of lack of alternatives. This is referred to as **passive coloniality**.

Colonial nesting is very different from **colonial roosting**. In many parts of the world, it is a common sight to see birds roosting together. In many European countries, the Starling *Sturnus vulgaris* forms colonial roosts, while in India it is a common sight to see crows, as well as mynahs and other relatives of the Starling, roosting together in trees, especially during the winter months (Figure 2.10). In fact, close to these massive roosts, one is confronted with a cacophony of sounds as the squabbling, noisy birds settle in and prepare for the night. This is another type of group behaviour with costs and benefits associated in a manner that is similar to group nesting. However, it is different from colonial nesting in a very fundamental way. A nesting colony is what is known as a **central place system** (Krebs and Davies 1984) – individuals belonging to the group must return to it regularly during the nesting season because it has eggs or chicks in the nest which require care and tending to. A roosting colony, on the other hand, is not a central place system in that an individual which has been a part of the group is not bound to return to it and is at liberty to go somewhere else and join another roost.

While in many countries in the temperate zone (as in the United Kingdom) heron colonies tend to be monospecific – nests of Grey Herons *Ardea cinerea* only – with several of them on a single tree, in places like India, colonies of heronry birds tend to be in mixed-species colonies and in a clump of trees or

FIGURE 2.10 A roost of migratory Rosy Starling *Pastor roseus* in India. (N. K. Tiwary)

even a single tree there may be nests of storks, ibises, spoonbills, cormorants, herons and egrets (Figure 3.1). Among Indian storks we know, for instance, that while Asian Openbill, Painted and the two adjutant storks are colonial nesters, others like Woolly-necked and Black-necked Storks are always solitary nesters. Biologists have attempted to theorise the possible advantages of **mixed-species colonies** as opposed to **single-species colonies** (Burger 1981).

Finally, there is the interesting question as to whether coloniality is **facultative** or **obligate**? Can individuals of a species nest in colonies in one situation while in a different place and context being solitary nesters? This is a fairly simple question but there seems to be no definite answer in the literature. Bildstein (1993) has discussed at length the nesting behaviour of the colonially nesting White Ibis *Eudocimus albus* over the course of several years with varying climatic conditions, and provides some interesting insights. The Painted Stork is almost always colonial, though the size of its colonies is highly variable and depends upon several factors including the extent of the substrate (tree canopy) available. While some of its colonies are recorded as being quite small, just two to three nesting pairs on a tree, other colonies can sometimes be large, with 20-plus nests. Nevertheless, one does occasionally come across single Painted Stork nests (Figure 2.11).

FIGURE 2.11 Single nests of Painted Stork are rare. (N. K. Tiwary)

Chapter 3

Painted Stork Colonies in India

Having laid a foundation for understanding how coloniality works in the previous chapter, here we will provide a glimpse of the enormous diversity of heronries in India, restricting ourselves to some of the most interesting and best-known examples. Most of the nesting colonies discussed here are of the mixed type – multi-species colonies with herons, egrets, cormorants, ibises, spoonbills, pelicans and other species of storks, coexisting on the same trees (Figure 3.1). Before we start examining these diverse types of colonies, we will briefly enumerate the different species of birds that coexist with Painted Storks at mixed colonies.

FIGURE 3.1 Mixed-species colony showing three different species of colonially nesting waterbirds: Painted Stork, Asian Openbill, Grey Heron; as well as at least two species of cormorant. (Paritosh Ahmed)

Birds nesting in mixed heronries

Herons and egrets belong to family Ardeidae and are medium to large wading birds, typically with long legs and toes, long bills and long necks. One characteristic of these birds, which distinguishes them from storks, is that their long necks are folded when flying. The common herons found at mixed heronries include Indian Pond Heron *Ardeola grayii* (Figure 3.2) and the similarly sized Cattle Egret *Bubulcus ibis*, Great Egret *Ardea albus* and Intermediate Egret *A. intermedia*, as well as the Little Egret *Egretta garzetta*, which is an all-white bird with black legs and yellow feet (or 'socks') (Figure 3.3). In western India, especially in the coastal areas of Gujarat, a related species the Western Reef-Heron *Egretta gularis* (Figure 3.4) exists alongside and often interbreeds with Little Egrets. Another common species

FIGURE 3.2 Pond Heron or Paddy Bird. (N. K. Tiwary)

FIGURE 3.3 Little Egret. Note the bright yellow feet characteristic of this species. (N. K. Tiwary)

at Indian heronries is the Black-crowned Night Heron *Nycticorax nycticorax* (Figure 3.5). Grey Heron *Ardea cinerea* is also sometimes found in mixed heronries (see Figure 3.1, extreme left corner).

Cormorants (Phalacrocoracidae) and darters (Anhingidae) have all-black plumage (again, see Figure 3.1). They have been described as 'foot-propelled pursuit divers' (Wanless and Harris 1991) and possess remarkable adaptations for underwater foraging, including a partially wettable plumage and a capacity to store large volumes of air in the lungs to facilitate longer dives. During a foraging bout, each dive is followed by a period on the surface for breathing and ingesting prey (Mahendiran and Urfi 2010). Three species of congeneric cormorants which coexist at heronries across India are the Little Cormorant *Microcarbo niger*, Indian Cormorant *Phalacrocorax fuscicollis* and Great Cormorant *P. carbo*. Cormorants and the closely related Oriental Darter *Anhinga melanogaster* are also found across India (Figures 3.6 and 3.7). Contrary to piscivorous waders, such as herons, egrets and storks, which are restricted to the littoral zone, cormorants are the main predators in the deeper-water regions of wetlands.

Members of the spoonbill and ibis family, Threskiornithidae, are also to be found at Indian heronries. The Eurasian Spoonbill *Platalea leucorodia* is a fairly common bird, slightly smaller than storks and all-white with a characteristic

FIGURE 3.4 Reef Heron. (N. K. Tiwary)

FIGURE 3.5 Night Heron. (N. K. Tiwary)

FIGURE 3.6 Great Cormorant (left) and Little Cormorant (right). (N. K. Tiwary)

spatula-shaped bill (Figure 3.8). Another member of the family, the Black-headed Ibis *Threskiornis melanocephalus* (also known as the Oriental White Ibis/ Indian White Ibis/Black-necked Ibis), is a common nester at heronries. It has a characteristic black head and a downcurved bill (Figure 3.9). Pelicans (genus *Pelecanus*) are large waterbirds belonging to family Pelecanidae. They are characterised by a long bill and a large throat pouch that is used when catching fish. In India, three species of pelicans are found but the commonest one, often nesting in mixed colonies in South India, is the Spot-billed Pelican (Figure 3.10). Asian Openbill Storks are also found at many heronries alongside Painted Storks (see Figure 3.1).

Resource partitioning in foraging has been studied in different heronry birds at Keoladeo Ghana National Park (Ishtiaq et al. 2010). With respect to nesting while several species coexist in the same trees, it appears that there is little interspecies competition for nesting sites or nesting material. Openbills make platform nests which are often utilised by Painted Storks, but there is little actual competition between these two species because their nesting seasons do not overlap. In northern India, Asian Openbills commence nesting in June, at the start of the monsoon season, while Painted Storks commence nesting from the end of August to September and hence there is some spacing between the two in a temporal sense (Figure 3.11). Egrets and

FIGURE 3.7 Darter also known as 'snake bird' catching fish. (N. K. Tiwary)

FIGURE 3.8 Spoonbill. (N. K. Tiwary)

FIGURE 3.9 Black-headed Ibis. (N. K. Tiwary)

FIGURE 3.10 Spot-billed Pelicans sitting atop trees in Kokkare Bellur. (Bharat Bhushan Sharma)

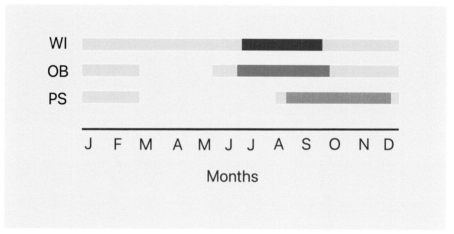

FIGURE 3.11 Figure showing how three species of similar-sized colonial waterbirds, Painted Stork (PS), Openbill Stork (OS) and Black-headed Ibis (BI), can coexist on the same substrate at Keoladeo Ghana National Park due to variations in their nesting period. Shaded areas denote presence, while coloured areas denote nesting activity. (Redrawn from Urfi 2011a)

herons are smaller species, existing alongside, but these birds tend to nest in the interior of trees on the smaller branches. The nesting season of the White Ibis coincides with the Painted Stork's, but they tend to form separate groups and build nests on separate parts of the substrate. Interestingly, Ali and Ripley

(1987) use the term 'mohalla' for such segregations in a mixed colony. It is a somewhat derogatory term used for local settlements of some communities across central Asia.

Types of heronries

Some heronries in India lie within protected areas while others exist in the wider countryside, where they are often vulnerable to destruction due to the activities of local people. In some cases, the nesting trees face the threat of being chopped down. Some heronries are situated in urban areas, in the middle of islands in ponds located in city parks and zoos or on tall trees in public gardens. Our observations over a wide range of heronries across India indicate that they fall into two broad types: colonies on tall trees and colonies on short trees surrounded by water (Figures 3.12 and 3.13, and see Figure 2.3). As discussed in the previous chapter, predation from the ground seems to have been a major factor in shaping colonial nesting behaviour in waterbirds. Among nests in the 'tall tree' category the common trees utilised are Silk Cotton Tree *Bombax ceiba*, which can grow up to 20 m or more in height, large *Ficus* trees and sometimes even *Eucalyptus* species. Such huge nesting trees can be in the middle of agricultural fields with a waterbody close by. It is also possible that a site may have a mixture of both 'island' colonies and 'tall tree' colonies. Delhi Zoo is a case in point. In years of normal rainfall, most storks build nests on islands in the main ponds of the zoo, which present a perfect island situation; but in years of exceptional rainfall, when much higher numbers of storks arrive at the zoo to nest, some tall Silk Cotton trees in the zoo grounds are also used. Some important Painted Stork nesting colonies in India that are discussed in various chapters in this book are described below.

Nesting colonies in urban sites

National Zoological Park (NZP) or Delhi Zoo: This well-known tourist spot in India's capital city is located on the western bank of the River Yamuna, which lies barely a kilometre away from the zoo (Figure 3.14). The park is located between two famous Mughal-era monuments, the Old Fort and Humayun's Tomb, and extends over 85 ha. Like any typical zoo, the NZP has a number of caged exhibits, but parts of the park are a mini-sanctuary for waterbirds. How this zoo became a nesting ground for the Painted Stork, and other species of heronry birds, is an interesting story described in Box 3.1.

FIGURE 3.12 Painted Stork nesting colonies can be located on very tall trees or on low clumps of trees surrounded by water. This photograph shows Painted Storks nesting on a tall tree in a village. Note that the nests are located at a height and so are safe from activities on the ground. (N. K. Tiwary)

FIGURE 3.13 In rare cases nesting colonies can be on offshore islands. Man Marodi Island, Gulf of Kutch, western India, where Painted Storks build nests quite close to the ground. (Hembha Vadher)

FIGURE 3.14 Location of Delhi Zoo in relation to the Yamuna. (Google Earth 2023)

About nine species of waterbird build their nests on clumps of *Prosopis juliflora* (mequite) trees with merged canopies in the zoo's ponds (see Table 7.1 and Figure 3.15). One of these ponds is located opposite the ramparts of the old fort, and the adjoining pond (also known as the pelican pond) lies adjacent to it. In some years – usually those in which the monsoon rains are

BOX 3.1 How a zoo became a waterbird sanctuary

How Painted Storks and other species of heronry birds began nesting on clumps of trees in ponds at Delhi Zoo is a fascinating story, the origins of which can be traced to the initial years of India's independence from British rule. But first let us look briefly at the history of zoos throughout the world and how they have evolved.

Zoological parks in some form or other have existed through the ages, though their objectives and purposes have varied. Early zoos were not meant for the public display of animals in the same manner as today's zoos are. In earlier times, animal exhibits served to signal the expanse of the ruler's kingdom and the wild animals found therein. The rulers built royal menageries in which they displayed birds and animals in ornate though wholly inappropriate and, most certainly, terribly uncomfortable enclosures.

In the age of exploration and discovery from the fifteenth to the nineteenth century, as European sailors and explorers travelled to hitherto unknown lands in Asia, Africa, the Americas and Australia, they brought back specimens of flora and fauna from lands they had conquered or visited. These specimens were often displayed in their home countries, leading to the creation of the first public zoos. The display arrangements in some of the earliest zoos in European cities were along the lines of geographical regions, with the fauna of Asia, Africa, the Americas and Australia all shown separately. The general public in European countries had little idea about the wealth of biodiversity across the globe, and when confronted with exotic and strange looking creatures exhibited in zoos they were fascinated and thrilled. Meanwhile, the zoos also continued to evolve. Regional settings in zoos began to be replaced by thematic settings; for instance, keeping the tiger or lion in one enclosure and their prey, deer, in a neighbouring one was an innovative way of educating the public about ecological interactions such as prey–predator relationships. With time, a new type of thinking came about: exhibition for the enjoyment of the visitor as well as the comfort of the live animal exhibit. While the zoos of yore displayed animals cooped up inside small cages, over time they began to be housed in large, open enclosures where they could move around and exercise their muscles. The cages, rods, mesh and other obstructions between the exhibit and the visitor were reduced, and moated enclosures became the norm. The animal exhibits looked less like they were held in captivity, as they were seen to be moving about freely.

After Independence in 1947, as a new country India strove to throw off the shackles of the past and develop as a modern nation-state. New institutions were created and there were efforts to harness atomic energy, launch satellites and explore the frontiers of space. Modern cities and hydroelectric projects were among the 'temples of modern India' being created, and there was also a need felt for a new, modern zoo to be constructed. India's first prime minister, Pandit Jawaharlal Nehru – a man with progressive, modern ideas – was emblematic of these developments. Conservation of wildlife and the environment was especially dear to him.

The leaders of newly independent India wanted India's national zoo to be modern in every respect. In 1952, the Indian Board for Wildlife created a committee for a zoo in New Delhi which would be the country's national zoo, hence the formal name, National Zoological Park. The facility would be built along modern lines with open, moated enclosures – a globally adopted concept originated by Carl Hagenbeck (1844–1913) of the Zoological Garden of Hamburg. Under this scheme there would be plenty of open space in the enclosure where the animals could move around. A suitable spot between the Old Fort and Humayun's Tomb was chosen as the location. It is said that Nehru took a personal interest in the project.

One of the unique features of the layout was that its exhibition areas and enclosures were to be located around interconnected ponds and canals, which would be narrow or wide at different points to accommodate various outdoor exhibits. A couple of such ponds were designed to showcase the waterbirds of India. Here, some mesquite trees were planted on islands and

Painted Storks at Delhi Zoo, with the ramparts of the Old Fort in background.
(N. K. Tiwary)

a few pinioned storks, cranes and other birds were released. The park was officially opened on 1 November 1959, but earlier in the year, probably in September, a small group of wild Painted Storks happened to fly around the zoo. They may have been scouting for a safe place to build nests. According to old records, about 50 to 60 birds descended upon the trees on an island in the pond facing the old fort. Although the birds did not build nests that year, a large contingent of about 400 Painted Storks landed in the zoo premises after the monsoons the following year and built their nests – and have been doing so every year since.

above average and also spread across a longer period, and large number of Painted Stork have flocked to the zoo for nesting – a smaller pond located well within the zoo grounds and a bit away from the two ponds is used. In years of exceptional rainfall, some tall trees which are not on islands (particularly some Silk Cotton trees in the vicinity of ponds) are also utilised for nesting, as previously mentioned.

Painted Storks started nesting in the zoo grounds in 1960 and 2021 marked over 70 years of continual nesting. Every year after the end of the summer monsoon (south-west monsoon) from mid-August to early September, Painted Storks start congregating at the zoo; by March or April they have all left, dispersing in the countryside around Delhi. While all these birds are no doubt opportunistic nesters, their annual visit to the zoo could be indicative of a shortfall of natural nesting habitats in the countryside (Urfi 1996, 1997), as well as an environment of safety which prevails in the zoo.

The Painted Stork population of Delhi Zoo has been studied sporadically, though in a fair amount of detail, over the past seven or so decades, chiefly in 1966–1971, 1988–1992 and 2004 onwards. First, Dr J. H. Desai and coworkers studied the general biology of the species: nesting ecology, diet and foraging, and growth and development of nestlings in the 1960s (see references to their work in other chapters). Thereafter, more or less coinciding with the popular interest in counting waterfowl in India generated by the Asian Waterfowl Census in the late 1980s, I began to undertake studies on populations and the colonisation patterns of Painted Stork. From 2000 onwards, an informal and loosely defined stork study group was formed at the University of Delhi with several MSc and PhD projects on various aspects of Painted Stork at the zoo and elsewhere in the Delhi region undertaken, including investigations into foraging ecology, sexual size dimorphism and its consequences, and inter-colony variations in nesting parameters. Recently, studies aimed at modelling nesting success in Painted Stork populations were completed (Tiwary and Urfi 2016a). Thus, the Painted Storks of Delhi Zoo probably afford the best example of a detailed and consistent study on a single population of a wild bird from anywhere in India.

The zoo authorities put food (dead fish) for the birds into the ponds. Although the food is primarily intended for the pinioned exhibits, all of the fish-eating birds, including Painted Storks, rush to grab this free meal. However, there are reasons to believe that the Painted Storks nesting in the zoo grounds hunt their own food and have little use for the fish put out in the ponds so far as feeding their chicks is concerned. One indication of this is that they soar on thermals out of the zoo to marshes associated with the River Yamuna to obtain food of the right type and size. It is widely held that wetlands associated with Okhla Barrage Bird Sanctuary, a few kilometres away, are their favourite foraging ground.

Bhavnagar city gardens: The city of Bhavnagar is located in Gujarat on the coast of the Gulf of Khambhat. There are three distinct areas in the city where different species (Table 7.1) of colonial waterbirds nest: the Old City area, Peile Gardens and its immediate vicinity, and the suburban areas of Krishnanagar and Takhteshwar (Parasharya and Naik 1990). About 16 different species of trees have been recorded as being used by Painted Storks for nesting. The birds make frequent foraging trips to both freshwater and coastal sites at varying distances from their breeding colonies in the city gardens.

Water storage reservoirs in the city of Mysuru: The city of Mysuru has several reservoirs – probably initially village ponds which were absorbed into the city as it expanded. One of them, important from the point of view of heronry birds, is the reservoir at Kokkare Halle (Figure 3.16), which attracts nesting Painted Storks and Pelicans to trees growing on its premises (Rahmani et al. 2016). Another important Painted Stork nesting colony is in Karanji reservoir, which covers over 40 hectares (ha) and is located on the eastern edge of the Mysore Zoo (formally known as Chamarajendra Zoological Gardens). It has nine artificial islands on which *Acacia arabica* and *Ficus* trees have been planted. Some of these trees, as well as those on the shore, are utilised by Painted Storks, Spot-billed Pelicans, and other species of colonial waterbirds for nesting (Table 7.1).

Colonies in marshes/protected areas

Keoladeo Ghana National Park: This 29 km^2 park is a UNESCO World Heritage site as well as a Ramsar site (Ali 1953; Ali and Vijayan 1983; Sankhala 1990). This used to be the only wintering ground for the western race of the

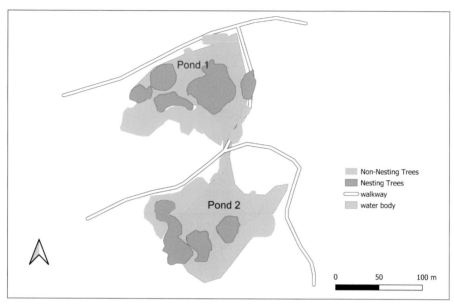

FIGURE 3.15 Sketch to show the two main ponds on the Delhi Zoo premises that are utilised by Painted Stork for nesting. These ponds which are a part of a series of interconnected ponds is an important feature of the zoo.

FIGURE 3.16 Kokkre Halle lake in Mysuru city. (A. J. Urfi)

endangered Siberian Crane *Leucogeranus leucogeranus*, which stopped visiting India altogether some decades ago. Meanwhile, Keoladeo is still an important refuge for migratory waterfowl and is also famous for its extensive heronries of Painted Storks and other birds.

Keoladeo was once the private hunting ground of the local ruler, the Maharaja of Bharatpur. Large stone slabs near the Keoladeo (also referred to as 'Keladevi') temple, from which the park derives its name, with figures of birds that had been shot (bags) etched on them, stand as testimony to earlier times when shooting and hunting was in vogue (Figure 3.17). A system of dykes and bunds was created to store the surplus waters of the River Gambhir, and rows of acacia trees were planted on the bunds and also on raised mounds in the middle of the marshes (Figure 3.18). Painted Storks and other heronry birds build their nesting colonies in these trees. Keoladeo was the site for a number of significant ornithological projects (including bird ringing) undertaken by the Bombay Natural History Society under the stewardship of the charismatic ornithologist Dr Salim Ali. Important field studies on avian migration and bird ringing, funded by international organisations, were conducted here. A long-term project on hydrobiology of the Keoladeo ecosystem (Ali and Vijayan 1983) continues to be the only authentic source of ecological data and baseline information about the site.

Sultanpur National Park: Sultanpur is located about 25 km to the southeast of Delhi in the Gurgaon district of Haryana, in a predominantly agricultural landscape crisscrossed by irrigation canals. The area was notified as a bird sanctuary by the Haryana State government in 1971. Later, an area of 13,727 ha, including as its core approximately 143 ha of low-lying marshes, was declared a national park (Figure 3.19). The dominant terrestrial vegetation in this area consists of trees of mesquite, Tamarind, Neem, *Acacia nilotica* and several types of grasses. The wetland has undergone several changes since the early 1970s when it became a popular birdwatching site, frequented by birdwatchers from Delhi. Peter Jackson, who was a Reuters correspondent based in Delhi in the 1970s (later chairman of several Species Survival Commissions for the International Union for Conservation of Nature or IUCN), is to be credited for having brought Sultanpur to prominence and lobbying to get it declared as a protected area. In the 1970s Sultanpur was just a shallow depression filled with brackish water which attracted a large number of birds, particularly Lesser Flamingo *Phoeniconaias minor*. However, as a part of its management policy, the state forest department embarked upon a tree planting programme during the 1980s. Several mounds were created in the lake and *A. nilotica*

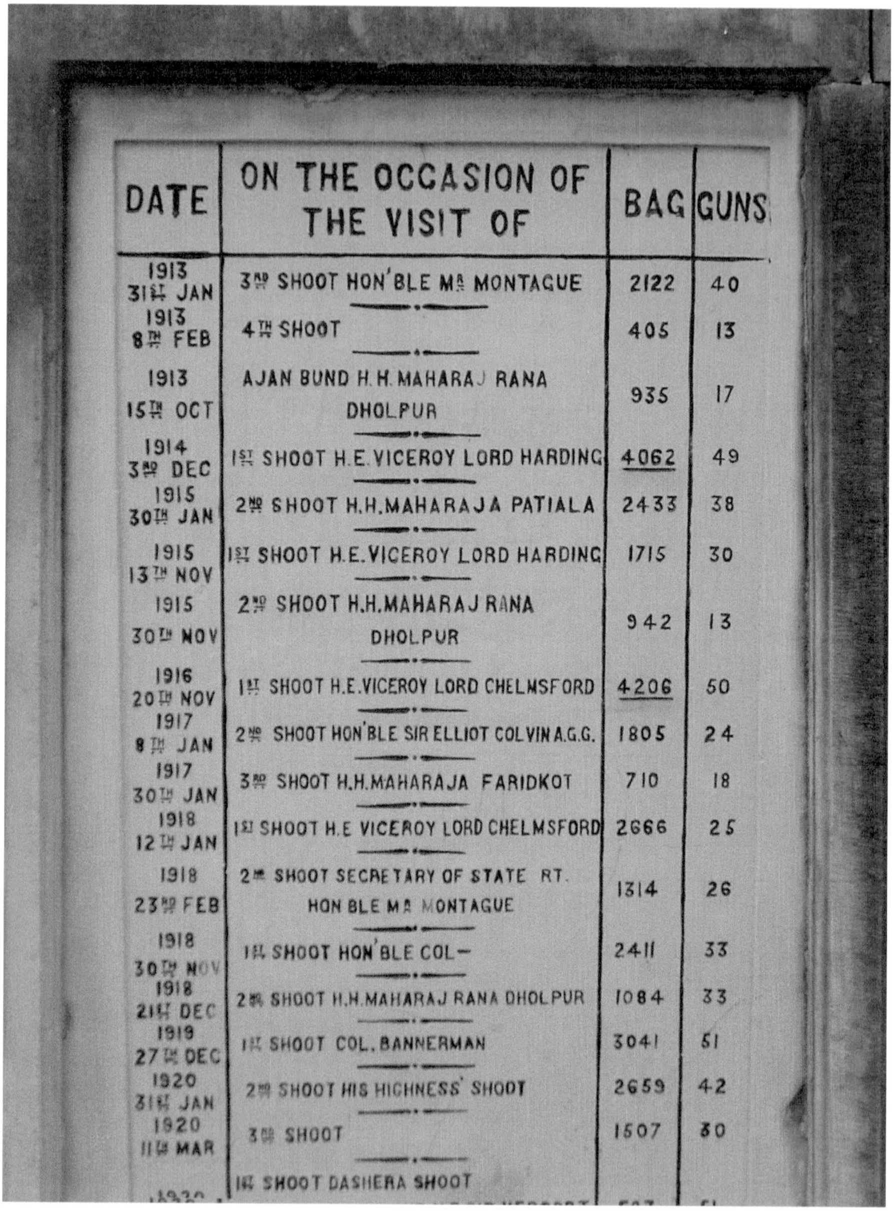

DATE	ON THE OCCASION OF THE VISIT OF	BAG	GUNS
1913 31ST JAN	3RD SHOOT HON'BLE MR MONTAGUE	2122	40
1913 8TH FEB	4TH SHOOT	405	13
1913 15TH OCT	AJAN BUND H.H.MAHARAJ RANA DHOLPUR	935	17
1914 3RD DEC	1ST SHOOT H.E.VICEROY LORD HARDING	4062	49
1915 30TH JAN	2ND SHOOT H.H.MAHARAJA PATIALA	2433	38
1915 13TH NOV	1ST SHOOT H.E.VICEROY LORD HARDING	1715	30
1915 30TH NOV	2ND SHOOT H.H.MAHARAJ RANA DHOLPUR	942	13
1916 20TH NOV	1ST SHOOT H.E.VICEROY LORD CHELMSFORD	4206	50
1917 8TH JAN	2ND SHOOT HON'BLE SIR ELLIOT COLVIN A.G.G.	1805	24
1917 30TH JAN	3RD SHOOT H.H.MAHARAJA FARIDKOT	710	18
1918 12TH JAN	1ST SHOOT H.E VICEROY LORD CHELMSFORD	2666	25
1918 23RD FEB	2ND SHOOT SECRETARY OF STATE RT. HON BLE MR MONTAGUE	1314	26
1918 30TH NOV	1ST SHOOT HON'BLE COL—	2411	33
1918 21ST DEC	2ND SHOOT H.H.MAHARAJ RANA DHOLPUR	1084	33
1919 27TH DEC	1ST SHOOT COL.BANNERMAN	3041	51
1920 31ST JAN	2ND SHOOT HIS HIGHNESS' SHOOT	2659	42
1920 11TH MAR	3RD SHOOT	1507	30
1920	1ST SHOOT DASHERA SHOOT		

FIGURE 3.17 Stone slabs marking the bags of duck shoots at Bharatpur during the time when the park was the private hunting preserve of the local ruler. (A. J. Urfi)

trees planted on them. There are now about 50 islands, including one large island (approx. 100 × 40 m) in the centre of the marsh. Painted Storks nest here, and on some smaller islands which have a canopy diameter up to 10 m, along with other species of colonial waterbirds.

FIGURE 3.18 Satellite picture of Keoladeo Ghana National Park at Bharatpur. The park is divided into a number of smaller marshes separated by embankments; Painted Storks and other heronry birds nesting in acacia trees planted on mounds in the marshes. The park is surrounded by agricultural fields. (Google Earth 2023)

FIGURE 3.19 A view of heronries at Sultanpur National Park, close to Delhi. (A. J. Urfi)

Painted Storks were first observed breeding in Sultanpur in 1993, when 40 nesting pairs were recorded. Thereafter, nesting was sporadic until around 2000, after which it became a regular event. Although the Sultanpur heronry has become well established, changes in the immediate environment of the park – such as loss of wetland habitats and increases in built-up areas, noise, traffic and pollution – are causes of concern. The original *jheel* system of which Sultanpur was a part seems to have been greatly affected by these developments (Gaston 1994; Poole 1994). Field studies have indicated that Painted Storks nesting in Sultanpur head out in the direction of the marshes alongside the Najafgarh drain, and they have also been observed flying in the direction of the Basai wetland and an unknown destination lying towards the west. All of these outlying areas, beyond the boundaries of the park, are undergoing transformation due to urbanisation and loss of wetlands (Urfi 2011a), which is bound to have an impact on the heronry birds nesting inside Sultanpur National Park.

Heronries in the countryside and villages

Kokkare Bellur: Popularly known as the 'Stork village', the site is situated about 80 km from the city of Bengaluru, in the Mandya district of the South Indian state of Karnataka (Rahmani et al. 2016). It is a typical dryland village in southern India with its cultivated and fallow fields, cactus hedges, and trees in the fields and villages. Kokkare Bellur is bounded on the south side by the Shimsha River, which is said to be where the principal foraging areas of Painted Storks, pelicans and other waterbirds lie, including flooded agricultural fields (Figure 3.20). In the village itself a large number of trees line the perimeter, surrounding the homes of the villagers, and these are used by colonial waterbirds for nesting. The Painted Stork and Spot-billed Pelican reside in the village from December to June each year (Figure 3.21). It seems that pelicans start their nesting slightly earlier than the Painted Storks, and the two species select different trees. In this two-tiered system, on the ground there is a village of humans and up in the trees there is a village of nesting birds, carrying on with their normal activities of breeding and feeding, seemingly oblivious to the activities below. It has been recorded that 'even in the midst of wedding celebrations, when loudspeakers blast music all over the village, the raucous sounds of the birds, the quiver of heavy wings and the clatter of stork bills continue unaffected' (Manu and Jolly 2000: 1).

It is widely assumed that storks and pelicans have been nesting here for a long time though precisely since when is not clear. Village legends put it at more than a hundred years. Interestingly, the village's name, Kokkare, is the

FIGURE 3.20 Painted Storks foraging in flooded agricultural fields in the vicinity of Kokkare Bellur. (Asad Rafi Rahmani)

FIGURE 3.21 A view of Kokkare Bellur village showing Painted Stork nesting colonies in the background. (A. R. Rahmani)

local name for stork, which bears out the long association. It is widely held that in 1864 when T. C. Jerdon (Jerdon 1864) made the following observation about a heronry in South India, but did not name it, he was probably referring to the village of Kokkare Bellur: 'I have visited one Pelicanry in the Carnatic, where the pelicans have (for ages I was told) built their rude nests, on rather low trees in the midst of a village, and seemed to care little for the close and constant proximity of human beings.' The Kokkare Bellur heronry remained unknown for a long time until it came to the notice of local birdwatchers in the 1970s, chiefly Neginhal (1977) who highlighted its significance.

Colonies in village ponds

Some of the commonest types of heronries encountered in India today are those located in the middle of village irrigation ponds and reservoirs, all across the country. In the state of Gujarat in western India, Seelaj, Traj (Urfi 2017) and Bhadalwadi tanks (Pande 2006) have good examples of such heronries. Typically, a large irrigation pond is filled with water during the monsoon. Trees (either acacia or *Prosopis*) are clustered on islands in the middle of the pond, or there is an island or several islands which are used for nesting by Painted Storks and other birds. In some cases, these ponds are areas of high biodiversity value; for instance, Traj irrigation pond has a small population of Mugger Crocodiles *Crocodylus palustris*, besides the heronry. In recent years, surveys in the Delhi region by the stork research group of Delhi University have resulted in documentation of Painted Stork colonies at Chhata, which is along the Delhi–Agra highway, close to the city of Mathura. The ramparts of an old Mughal fort in its vicinity makes this a very attractive site. Painted Stork colonies are located at Chhata and Khanpur villages also (Tiwary et al. 2014).

   Koonthankulam Bird Sanctuary: Located in Naguneri Taluka of Tirunelveli, about 20 km from the town of Tirunelveli, this rain and river-fed freshwater reservoir is an important nesting ground for both Spot-billed Pelicans and Painted Storks (Figure 3.22). The pelicanry is said to have existed for 200 years or more and has been much written about, both historically (Rhenius 1907; Webb-Peploe 1945; Wilkinson 1961) and later (Nagulu and Rao 1983). The site is visited by all manner of waterbirds and is a popular place for student tours (Rahmani et al. 2016; BirdLife International 2023c).

Offshore colonies

Painted Stork nesting colonies can also be found on islands in the sea. Man-Marodi Island in the Gulf of Kutch, off the coast of Gujarat, first

FIGURE 3.22 Koonthagulam Bird Sanctuary. (M. Mahendiran)

documented by this author (Urfi 2003a), is a good example (see Figure 3.13). Here, Painted Storks nest very close to the ground on cactus plants. The types of predators and threats the nesting birds face at this site are largely unknown and this aspect needs to be studied. Colonies in mangrove forests on either the east or west coast, if any, are poorly documented.

Chapter 4

Nesting

The timing of bird reproduction is an important issue in ornithology. In this context, ecology textbooks refer to two factors of great significance: the 'proximate' and 'ultimate' factors. The ultimate factors governing reproduction in birds relate to annual food cycles and are responsible for the huge variability of nesting times all across different areas, which will be discussed in detail in Chapter 6, in relation to the discussion on foraging ecology.

By the time an individual has reached a stage when it is on the lookout for a mate and ready to commence breeding, several physiological processes have been underway in its body for some time. Some of these processes are maturation and growth of germ cells, mobilisation of resources from body tissues towards gonad development and formation of nutrient reserves such as yolk in eggs to sustain embryos. These physiological processes are triggered by environmental factors including seasonal changes in temperature, humidity, day length and so on, known as the proximate factors, which, by a complex series of biochemical reactions, begin the process of preparations for reproduction in the bird's body. Perhaps the most obvious of these is the development of gonads, which enlarge only at the time of reproduction and regress soon afterwards. This is linked to changes in body weight and aerodynamics constraining flying and marks a contrast to the situation in mammals, in which the gonads remain more or less fully developed throughout the year.

Painted Storks are colonial nesters but not communal or group roosters. This means that in the non-breeding season they are widely scattered all across the countryside, looking for food in wetlands, agricultural fields and other suitable habitats. It is only when the breeding period starts, August to September in North India, that they start flocking to their traditional nesting sites. So, in August, when they arrive at Delhi Zoo, Painted Storks have a fresh glistening look and a spotless white plumage. The bill is bright waxy yellow and the head is fully bald (Figure 4.1). This is due to hormonal secretion which

FIGURE 4.1 At the commencement of the nesting season Painted Storks have glistening, waxy yellow bills and a broad bald patch on the head. (N. K. Tiwary)

is enhanced at the time of reproduction (Shah et al. 1977a&b). However, as time passes the plumage gets soiled and dirty.

The courtship patterns of Painted Storks were studied in detail by Philip Kahl in the 1970s, including observations made at several places in India (Kahl 1970). Using cinematography, Kahl recorded nesting behaviour and replayed the footage at slow speeds to find out what was happening at the time of courtship (Kahl 1971b). According to him, upon arrival at the colony site the birds at first simply stand on tree limbs in groups of unattached males and females (Figure 4.2). Then adult males attempt to take possession of a suitable territory, which could be a nest used in previous years or, in a few cases, a site where a new nest is built from a scratch. Sometimes, they use nests of other species, such as the Asian Openbill, which build nests on the same trees. In northern India, Asian Openbills commence nesting in June and by the time Painted Storks arrive most of their nests are vacant.

A male Painted Stork attracts a female by advertising his presence, though with such subtlety and finesse that his actions hardly look like a proper display (Figure 4.3). A lone male may appear to be doing nothing except preening himself and rearranging twigs in the nest, but in actuality he may well be

FIGURE 4.2 Painted Storks settling in at the nesting colony at the beginning of the nesting season. (N. K. Tiwary)

FIGURE 4.3 Preening is a form of territorial display. (N. K. Tiwary)

displaying (Kahl 1971b). Should another male encroach upon an occupied territory, the incomer is rudely driven off by a powerful jab of the bill. Some interlopers fight back, and sometimes when conflicts escalate the intruder may succeed in usurping the territory. Colonial waterbirds are quite aggressive during territorial displays and at the beginning of the breeding season there is generally much action to be seen at a colony (Figure 4.4). A female drawn to a displaying male is initially attacked as if she were a rival male. But instead of retaliating, she adopts a meek and submissive posture, her head held low, wings spread wide and bill gaping (Kahl 1971b). The male may begin to see her as a potential mate and if a pair bond develops between them, they immediately embark upon a hectic schedule of nest-building and mating (Figure 4.5).

According to Ali and Ripley (1987: 24), the breeding season of Painted Stork is 'variable, dependent on monsoon conditions. Normally August to October in North India; November to March in the south; March to April in Ceylon [Sri Lanka]. In drought years breeding may be skipped altogether.' The reasons for the huge variability of nesting times across India is discussed in detail in Chapter 6, so here we will take note of some interesting and unique nesting behaviour patterns in the Painted Stork.

Age at first breeding: Being large birds, the age of first reproduction in the Painted Stork could be expected to be at least a three or four years. The only

FIGURE 4.4 Action at the beginning of the nesting season includes behaviours such as greetings, displays and fights. (N. K. Tiwary)

FIGURE 4.5 A couple of weeks after arrival at the colony site Painted Stork have established territories and the early nesters are already incubating eggs. (N. K. Tiwary)

authentic information on record is from some colour-marked birds at the Delhi Zoo colony. In 1968, ten nestlings of known age were banded and released into the wild (Desai 1971). However, as the birds had become tame they did not leave the zoo premises. In September 1972, three of these colour-marked birds paired and mated with wild storks in the zoo colony, and it can be deduced from this that Painted Storks become sexually mature and breed for the first time at approximately four years of age.

Nest-building: The nests of Painted Stork, essentially 'platform nests' (Figure 4.6), are a jumble of twigs placed in the crowns of trees. Construction takes four to eight days. Nests built from scratch are rare, as those made in the previous year or abandoned by other birds are generally utilised.

Lining nest with green material: At the time of nest-building, Painted Storks collect fresh leafy twigs for lining their nests (Figure 4.7, left). At Delhi Zoo they have been observed pulling green branches of mesquite or acacia off with their bills. Some other species of colonial waterbirds have been recorded doing the same (Figure 4.7, right). Given that this is such a common behaviour among colonially nesting waterbirds, the question is what purpose does it serve? A study on Wood Storks (Rodgers et al. 1988) examined a number of hypotheses concerning the role of green vegetation, including concealment of eggs or young nestlings from predators, chiefly Raccoon *Procyon lotor* and Fish Crow

FIGURE 4.6 The platform nests of Painted Storks are always located on the outer substrate, making it easy for the birds to take off and land. (N. K. Tiwary)

FIGURE 4.7 Carrying green material to line their nests with in their beaks is a common practice of colonially nesting storks. (Left) Painted Stork. (Right) Asian Openbill carrying freshly plucked branches. (N. K. Tiwary)

Corvus ossifragus, and providing shade. However, nests are usually not left unguarded as one adult stork is generally present when eggs or hatchlings are in the nest. Furthermore, the nest contents are seldom covered by greenery so shading does not appear to be the case. It does not seem to be as important for storks as it is for some other species. A possible role for the greenery could be aiding nest sanitation since nestlings regularly defecate in the nest. However, since in the Wood Stork greenery deposition is initiated prior to egg deposition and declines subsequently, this is unlikely to be the primary factor. In fact, the greenery actually tends to prevent guano from passing through the nest twigs, which reduces the plausibility of the sanitation hypothesis.

Perhaps use of greenery in nests helps in repelling ectoparasites via release of secondary compounds during the drying or decay of plant material. European Starlings are reported to choose plants for nest-building whose volatile compounds are likely to inhibit arthropod hatching and bacterial growth. Rodgers et al. (1988) suggest that some of the vegetation used by Wood Storks does support this hypothesis, especially the aromatic or resinous species. However, Wood Stork nests are heavily infested by dermestid larvae, biting lice and mites, against which the vegetation used by storks has no effect. Significantly, in the different colonies studied the use of vegetation was more a reflection of what was easily available in any particular locality. It is possible, then, that green material in nest lining may serve more than one function. For instance, while one advantage could be that it helps to retain food regurgitated by parents, especially in the case of food boluses, it may also have a role to play in terms of maintaining the pair bond as well as upholding the structural integrity of the nest. Through a series of experiments, it was demonstrated that the green material used for lining could have a role to play in insulating the nest and reducing the energetic cost of incubation. Greenery, by itself or in combination with guano, may also help in plugging holes in nests with a twiggy, porous structure. Whatever the reason, lining the nest with greenery is certainly a very prominent part of nest-building in Painted Stork and other species.

Dealing with heat – shielding young, regurgitating water and urinating on legs: Storks live in warm climates throughout the world (exceptions being some *Ciconia* species; the White Stork, for example, nests in Europe but migrates to Asia and Africa in winter). Not surprisingly, therefore, storks exhibit behaviours that help in countering excess heat in their environment. Unlike some other species of birds, storks do not indulge in gular fluttering (rapid movement of neck muscles, with bill open, to promote heat loss).

A typical stork behaviour observed at nest is shielding the young from direct sunlight by spreading the wings half open but bent at the wrists to form a shield

(Figure 4.8), also called the delta wing pose. Much has been written about this pose which apparently serves the twin purposes of shielding the nestlings from direct sunlight and also sunning, which is helpful in restoring the curve to feathers after soaring has left them misshapen (Kahl 1971a). As the sun shines upon the nest from different angles during the course of a day, the parent birds shift their position accordingly. This idea has not been tested, though it is recorded in anecdotes. For instance, according to a report by Abraham (1973):

> The villagers assure me that the parent birds stand on eastern side of the nest in the forenoon and on western side of the nest in the afternoon to protect their young ones from the sun.... However, even the casual visitor cannot but notice the way the Painted Stork protects the young ones from the sun by spreading its broad and long wings over the young ones as a sort of improvised roof.

To bring down their own body temperatures in hot weather storks use a technique known as 'urohydrosis', which involves excreting on their legs (Figure 4.9; see Box 1.1). Defecating directly onto the legs enables heat loss due to evaporation (Kahl 1963). To cool down the nestlings in hot weather, Painted Stork sometimes regurgitate water on the heads of chicks. The water is scooped up from the nearest waterbody (Figure 4.10).

FIGURE 4.8 Painted Stork shielding its chicks from the sun by spreading its wings. (Paritosh Ahmed)

FIGURE 4.9 Storks urinate on their legs to help them keep cool. (Paritosh Ahmed)

FIGURE 4.10 Painted Stork feeding chicks away from the nest on the ground. Note the typical begging posture of the chicks. (N. K. Tiwary)

Climbing on thermals and circling in the sky: It is a common sight to see Painted Storks circling around the colony in a column in the early hours of the day. Many of the adult birds are getting ready to go on a food-finding mission and they use thermal columns of rising air, heated by the sun-warmed ground, to gain altitude – an energy conservation strategy adopted by many large-winged birds such as vultures (Pennycuick 1972).

Non-breeding individuals in a colony: Not all of the birds that congregate at the nesting site at the beginning of the nesting season will be able to success-fully breed. Suppose ten active nests were counted at a colony, assuming that each nest had two adults in attendance, the estimated roost count would be expected to be 20. It turns out that the answer is not that simple: at the beginning of the nesting season the number would be greater than 20 and towards the end of the nesting season it would be far fewer than 20, but never exactly 20. Careful counts have revealed that the surplus of non-nesting birds at the start of breeding is about 20–30% of the population. Since there have not been any studies using ringed or marked birds, we do not know anything about these surplus birds.

For many colonial waterbirds the nesting site also serves as a roosting site, so the occurrence of additional individuals may be coincidental or related to opportunities for obtaining food while chicks in the colony are being fed. However, Painted Storks congregate only during the nesting period, so this explanation is unlikely. The surplus individuals could be immature individuals attempting to seek extra-pair copulations (Westneat et al. 1990).

Helping behaviour has been observed in several birds, where closely related individuals help adults to raise young. So, perhaps, some of these apparently extra individuals are genetically related to the breeding pair and hang around the nest as helpers. In seabirds, nonbreeding individuals, both immature and adult, may be permanent residents within a colony, defending territories, constructing rudimentary nests and even assisting in raising young. But in the case of Painted Stork we do not have a clear-cut answer yet. In any case, the parents have to spend much of their time being vigilant and protective and defending their nests from other birds as well as taking care of the chicks – and so a little help, maybe from closely related individuals, is what they need (Figure 4.11). Cases of polygyny have been recorded in some ardeids. In a study in Japan, polygynous trios were recorded in 7.7% of Cattle Egret nests (Fujioka 1986). Several cases of polygyny were observed in the population of Asian Openbills nesting at Raiganj Wildlife Sanctuary in India (Datta and Pal 1995). Thus, storks, which have always been found to breed monogamously previously, may actually be

FIGURE 4.11 Chicks need to be guarded from danger so in the early stages of their growth at least one parent always stays at the nest. (N. K. Tiwary)

more polygamous than previously believed, though this is another subject which needs to be studied in detail.

Growth of nestlings

The criteria for classifying the young of any species as being 'altricial' or 'precocial' are multiple, so in reality there is an 'altricial–precocial spectrum' where that species (its young) can be placed (Starck and Ricklefs 1998). Nice (1962) attempted to clarify the picture by describing in detail the different diagnostic features of hatchlings for a variety of bird species so that they can be placed in discrete developmental categories, such as: Plumage of hatchlings (none, with down, with contour feathers); Eyes (open or closed); Nest attendance (stay in the nest, stay in nest area, leave the area); and Type of parental care (none, brooding, food showing, parental feeding). According to these criteria, Painted Stork nestlings are semi-altricial-1 because they are covered with down at the time of hatching, have open eyes, stay in the nest until they are quite grown and are fed entirely by their parents during the nesting period (Desai et al. 1977; Shah et al. 1977a&b). The Wood Stork is placed in the same category and probably some other storks too.

The details of eggs and clutch-size variations and post-hatching growth patterns in Painted Stork have been reviewed in Urfi (2011a). Painted Stork nestlings are extremely noisy and, as Rhenius (1907) observed, '[t]he young of the Painted Stork may be a pretty bird with his pencilled plumage, but he is a noisy brute and seems to spend most of his time trying to let everyone for miles around know how hungry he is.' Painted Stork nestlings grow extremely fast and require enormous quantities of food, and for much of the nestling phase the parent birds are kept busy collecting fish and regurgitating their crop contents in the nest, which are gobbled up by the chicks. Starck and Ricklefs (1998) estimated the growth parameters for a number of birds, including five species of stork. For the Wood Stork with an initial body mass = 2,500 g, the value of K_G (Gompertz function) = 0.085, and K_L (adjusted to Logistic Growth Function) = 0.125. Compared with a precocial, large North American wader, the Sandhill Crane *Antigone canadensis*, whose chick at the time of hatching weighs about 5,118 g, slightly more than double the weight of a Wood Stork nestling, K_G = 0.032 and K_L = 0.047. So, growth rates are very different for altricial and precocial birds.

In the literature, the ages of Painted Stork nestlings have been defined in loose terms, using such vague expressions as 'freshly hatched chicks', 'very young ones', 'considerably older chicks' and such like. These sorts of vague terms are not very useful if one wants to discover exactly when the nesting period started at any given site. Having a good idea of nestling age makes it possible to pinpoint the initiation of the nesting period with reasonable accuracy and this information can be extremely useful in modelling nesting success. Hence, there is a need to elaborate upon the different growth stages of nestlings in the field. What Painted Stork nestlings look like at different stages of their growth is illustrated in Box 4.1.

Physical changes continue over the first two years. During the first year the plumage starts turning white and eyes become deep straw-yellow. By 16 months the birds have become voiceless, like adults (Desai et al. 1977). In the second year the bird acquires a plumage almost like the adult's but the characteristic black band across the breast is not fully developed and overall, the body plumage is lighter than in a mature adult. The light yellow of face and bill becomes brighter and the inner secondaries of subadults develop the bright rosy-pink colour characteristic of adults.

Modelling nesting success

Nesting success (NS) is an important fitness parameter. The latest methods aimed at estimating NS, incorporating ideas proposed by Mayfield (1961, 1975), tend to consider an individual nest over the entire span of the nesting period, during which time it is exposed to the vagaries of the natural world. Daily Mortality Rate (DMR) is estimated for each day that the nest is active,

BOX 4.1 A synopsis of the development of nestlings and diagnostic morphological features of the Painted Stork at different ages

(source: This box is based on information reviewed in Urfi 2011a).

1. Just hatched to 24 hours post hatching – weight approx. 55 g

Plumage and morphology: Entire body sparsely covered with light grey down feathers (prosoptiles) which is thickest on dorsal surface of the wings. Ventral surface of wings almost naked. Eyes open and black. Tip of bill yellow.

Just hatched. Note eyes are closed and bill has a yellow spot at its tip. Less than 24 hours post hatching. (N. K. Tiwary)

Behaviour: Hatchling appears feeble and weak, movements uncoordinated, unable to stand on its feet or lift the disproportionately large head properly. Most time spent sleeping in the nest. Occasionally emits loud squeaks. Hatchling unable to pick up food in the nest.

2. Very young nestling, 1–15 days post hatching (DPH; 1–2 weeks) – weight range 144–526 g

Plumage and morphology: Down on dorsal side becomes denser. By 11–15 DPH, prosoptiles replaced with dense woolly, white, second generation of down (mesoptiles). Bill turning black but has a yellow tip.

Very young nestling, 1–15 DPH. (Paritosh Ahmed)

Behaviour: Nestlings make loud squeaking noises. From 6–10 DPH, they are capable of moving head and neck at will. Jabbing movements of bill seen while picking up fish from nest. Nestlings can make short movements by standing in nest on tarsi. Muscle tone improves and they start exhibiting coordinated actions like preening and cleaning of wings. Tendency to bask in sun with wings spread out. May make up–down display movements. Able to eat fish about 7–9 cm long, which the parents regurgitate in the nest.

3. 16–30 DPH (3–4 weeks) – weight range. 1,073–1,249 g

Early stage, 16–30 DPH. (N. K. Tiwary)

Plumage and morphology: Nestlings have soft, dense woolly white down. This is referred to as cotton ball/snowball stage. Primary feather stubs are clearly discernible. Bill black, eye colour black or deep amethyst. Overall, appearance considerably bigger than at hatching, fluffy, white with a jet black head and bill; some black on body due to development of some black retrices and remiges.

Behaviour: Nestling considerably noisier, resorting to loud cries when hungry or at approach of parents bringing food. Up–down display greeting towards parents bringing food. Nestlings sometimes left alone in the nest. Muscle tone has improved considerably and body movements perfectly coordinated. Nestlings frequently flap and exercise wings; able to stand on tarsi for longer periods of time, but not yet on their feet. Threat attitude sometimes observed among nestlings.

4. 31–45 DPH (5–6 weeks) – weight range. 1,599–1,960 g

31–45 DPH. (N. K. Tiwary)

Plumage and morphology: This stage sometimes referred to as 'turkey stage', probably due to large size (half adult weight) and movements of nestlings. Plumage still generally white, but feathers on the back slowly changing to black. Big patches of black are visible on the dorsal part but towards 41–45 DPH head and neck feathers begin to turn smoky grey. Yellow tip of black bill looks slightly larger.

Behaviour: Nestlings still very noisy. For the first time the nestling can stand up and walk on its feet in the nest. Movements are quite unrestricted and nestlings are seen exercising wings and attempting to jump. Threat displays toward nestlings in the neighbouring nests are sometimes seen.

5. 46–60 DPH (7–8 weeks), weight range. 2,230–2,650 g (3/4 adult size)

46–60 DPH. (N. K. Tiwary)

Plumage and morphology: Head and neck plumage turns smoky grey and young look grey overall. Nestlings 3/4 adult size. The black bill starts becoming lighter in colour. Remiges and retrices are well developed; eye colour changes from amethyst to brown.

Behaviour: Motor activity is considerably heightened and includes flapping of wings and hopping. Sometimes small flights from the nest to nearby branches; 'bouncing off' the nest is commonly observed. As they approach 60 DPH nestlings start flying around in the colony. Tree-to-tree flights are rare. Still noisy but over time a bass quality of voice replaces the shrill calls of younger nestlings.

6. 61–70 DPH (9–10 weeks), weight range. 2,833–3,008 g

61–70 DPH. (N. K. Tiwary)

Plumage and morphology: Nestlings almost adult size and fully fledged. Face and bill considerably lighter in colour and they are smoky grey overall.

Behaviour: Though still dependant on food brought by parents, some are seen foraging along with adults in vicinity of nests. Local flights have become common. Important developments are much reduced voice and clattering of mandibles is evident.

7. 71 DPH (11 weeks) to 9 months.

71 DPH. (Paritosh Ahmed)

Plumage and morphology: Now fully grown, nestling looking more like an adult. Head and neck still smoky grey but retrices and remiges have turned black. Face and bill turn light yellow. Bill slightly decurved at the tip though less than in adults.

Behaviour: Flights around the colony are common. Parents still bring food to nest, though nestlings also forage locally. After 85 DPH, the grey young have become independent in all respects and feeding by parents has ceased. The voice is much reduced and clattering of mandibles common.

including the period of incubation and fledgling phases. The resultant estimate of daily survival rate (DSR), the inverse of DMR, is defined as the probability of a nest producing at least one surviving young of a certain (biologically relevant) age, which is 45 days post hatching (DPH) in the case of Painted Stork.

With the development of computing software, it is now easy not only to estimate DSR but also model it by accounting for the host of biotic and abiotic factors which are likely to influence the DMR of a nest. The field study involves repeated visits to the nest to monitor its status, which on a given day can be extinct (0) or alive (1). There can be several factors (covariates) that influence the status of a nest and it is for the investigator to see which ones can be reliably quantified and included in modelling exercises. Some of these covariates are discussed here in the context of the Painted Stork.

Firstly, nest predation is an important consideration. Scavenging of nestlings carcasses by raptors has been recorded as a significant cause of nest failure in related Ciconiiform species (Frederick and Collopy 1989) and also in context of Painted Stork (Naoroji 1990; Urfi 2010a). Weather conditions can also play an extremely important role. For instance, inclement weather during some phases of the nesting period can lead to mortalities of vulnerable nestlings (Dawson and Bortolotti 2000; McCreedy and van Riper 2015). While temperature-induced scarcity of fish in wetlands is also known to influence the nesting of piscivorous waders (Frederick and Loftus 1993), the quantity of rainfall also strongly affects reproductive output by regulating prey availability (Djerdali et al. 2008). Thus, loss of seasonal wetlands in the vicinity of colony sites can lead to delayed and poor nesting (Frederick et al. 2009). Seasonal or annual variation in prey availability can also affect nest survival and population dynamics of the nesting birds (Beerens et al. 2011).

In the traditional cost–benefit analysis framework of avian coloniality (Wittenburger and Hunt 1985), it has been argued that the benefit accruing to individuals building nests in high-density patches is likely to influence nest survival rates, and hence both colony size and nest density are important. Inundation following periods of prolonged rainfall results in island-like nesting trees offering enhanced protection from mammalian and reptilian ground predators. Studies on a variety of different bird species have examined nest survival in the context of colony size (Szostek et al. 2014), weather conditions (Dinsmore et al. 2002; Collister and Wilson 2007), dry conditions and cooler temperatures (Dreitz et al. 2012), daily precipitation rates and even wind speed.

Studies at Delhi Zoo and other North Indian sites

In a field study we explored the role of various factors that are likely to influence nest survival in four different Painted Stork colonies in northern India (Tiwary and Urfi 2016a). Daily survival rates related to Painted Stork nests from the four nesting colonies (Delhi Zoo, Keoladeo National Park, Chhata and Khanpur) were modelled as a function of multiple covariates, the list of which is provided in Table 4.1. Besides nest-specific and colony-specific parameters, it included

TABLE 4.1 Model notations for different covariates used in the competing models. DSR, daily survival rate (Source: Tiwary and Urfi 2016a)

Model	Notation
1. Single estimate of daily survival	$S_{(.)}$
2. Variation in DSR with nesting season	S_{DATE}
3. Variation in DSR within sites	S_{SITE}
4. Annual variation in nest survival	S_{YEAR}
5. Effect of extent of wetland around the nesting colony	S_{WETL}
6. Effect of colony type	S_{CTYP}
7. Effect of nest age	S_{NAGE}
8. Effect of nest height	S_{NHGT}
9. Effect of nearest neighbour distance	S_{NNBD}
10. Effect of nest density	S_{DNES}
11. Effect of colony size	S_{CSIZ}
12. Effect of nest age and a linear trend	$S_{NAGE} + T$
13. Effect of nest age and a quadratic trend	$S_{NAGE} + T_T$
14. Best model plus maximum daily temperature effect	$S_{BEST} + T_{MAX}$
15. Best model plus minimum daily temperature effect	$S_{BEST} + T_{MIN}$

year, daily maximum and minimum temperatures, and size of foraging wetlands in their vicinity. Considerable interannual variations in nest survival rate are known among birds (Dinsmore et al. 2002). Since variations in rainfall are likely to influence the food production cycles of an exclusively piscivorous bird like the Painted Stork, variation in a specific year could result in a 'year effect' that accounts for variations not caused by nest-site characteristics alone. Similarly, ambient temperature could have an influence on DMR in that in the winter months (particularly December and January) minimum temperatures can drop quite low (1–4°C and sometimes even lower) in North India, causing deaths of chicks due to hypothermia.

A total of 1,095 Painted Stork nests (606 in 2013–14 and 489 in 2014–15) were monitored at intervals of 4–7 days during a study spanning two breeding seasons. Repeated sightings of the same nest and monitoring its fate (0/1) is an essential feature of this approach, which is based on the idea of 'mark and recapture'.

We found that nest survival for Painted Storks is significantly affected by environmental factors like rainfall and daily minimum temperature, along with colony-specific parameters such as nest density. Our main findings (see Table 4.2) can be summarised as follows. Nest survival was found to be strongly dependent upon nest age in that nests built early in the season had a higher

TABLE 4.2 Model selections for Painted Stork nest survival data for nests monitored during the 202 days (including egg-laying, incubation and fledging stage) of nesting seasons 2013–14 and 2014–15 (combined) for four different Painted Stork colonies studied in the Delhi region. Models are ranked according to the Akaike's information criterion (AIC). Best Models (Δ AIC < 2) are shown in bold type. (Source: Tiwary and Urfi 2016a)

Model	Δ AIC
{S (DATE+NAGE+TT+YEAR+SITE+T_{MIN})}	**0.00**
{S (DATE+NAGE+TT+YEAR+SITE)}	**0.04**
{S (DATE+TT+NAGE+YEAR+ T_{MIN})}	**0.70**
{S (DATE+NAGE+TT+YEAR+WETL)}	**1.04**
{S (DATE+TT+NAGE+YEAR)}	**1.37**
{S (DATE+NAGE+TT+YEAR+DNES)}	**1.51**
{S (DATE+NAGE+YEAR)}	**1.79**
{S (DATE+TT+NAGE+NHGT)}	2.03
{S (DATE+TT+NAGE)}	2.30
{S (DATE+NAGE+TT+NNBD)}	2.38
{S (DATE+NAGE+TT+DNES)}	2.53
{S (DATE+TT+NAGE+CTYP)}	2.64
{S (DATE+NAGE+TT+SITE)}	3.56
{S (DATE+NAGE+TT+CSIZ)}	3.97
{S (DATE+TT)}	4.51

survival probability ($\beta_{Nest\ age}$ = 1.13, SE = 0.16). Daily survival rate was higher for nests located in high nest density colonies ($\beta_{Nest\ density}$ = 0.64, SE = 0.04). The additive effect of daily minimum temperature on the nest survival rate was positive ($\beta_{Temp.\ min}$ = 1.21, SE = 0.26), indicating that lower temperatures in winter negatively influenced daily survival rate.

The nest initiation date (the date on which a nest becomes active, back-calculated from observations of its status on the day it was first observed) is crucial here in the sense that the variations in nest timing create asynchrony and so successive cohorts can experience different environmental conditions (McCreedy and van Riper 2015). Thus, very late nests can be subjected to extreme weather conditions, resulting in elevated failure rates (Dawson and Bortolotti 2000). However, early nests may benefit from optimal weather conditions, and these may often be initiated by individuals that are in better condition compared with late-season nesters. Previous studies on the Delhi Zoo Painted Stork population suggest that birds mating early in the season tend to be those that have larger body size (a surrogate of condition) compared with those mating later (Urfi and Kalam 2006). This means that nests built early will have nestlings in an advanced stage of development later in the season when environmental conditions change and become adverse. These quite grown-up nestlings may be better equipped for survival. For instance, later nestlings may not have adequate protective feathering on their bodies to prevent hypothermia and therefore may be at a greater risk when temperatures drop in December and January.

While not all cases of nest losses in this study could be recorded, some general observations made imply that several factors could be involved. During the winter months, especially December and January, raptors such as Imperial Eagle *Aquila heliaca*, Steppe Eagle *Aquila nipalensis* and Crested Serpent Eagle *Spilornis cheela* were observed feeding on carcasses of both adult and nestling Painted Storks at some sites other than the zoo (Naoroji 1990; Urfi 2010a). These mortalities were possibly caused by hypothermia because for several days in December and January the minimum temperatures dropped as low as 4°C, accompanied by chilly winds. In contrast, in August and September, when daily maximum temperatures can reach up to 45°C, Painted Stork parents employ behavioural adaptations such as using their wings to create shade for nestlings or regurgitating water to cool the nest and nestlings during the hottest hours of the day. Unsurprisingly, maximum temperature does not emerge as a significant covariate in the nest survival models.

Temperature impacts different species in different ways (Cox et al. 2013). For instance, in a study of DSR, for some species drought conditions and cooler temperature can increase nest survival rates because these extreme conditions influence predation rates (Dreitz et al. 2012). Nesting of some wading birds is also known to be affected by temperature-induced fluctuations in fish availability. Even a marginal decline in temperature beyond a certain threshold can significantly affect fish density in wetlands (Frederick and Loftus 1993). The significance of minimum temperature in our study raises the question of whether nest survival rates are likely to be higher at those sites where the climate is benign and temperatures do not drop as low as in North India. This is of particular relevance for those nesting sites that are located in South India and Sri Lanka, where ambient temperatures rarely drop below 16°C at any time during the year (Urfi 2011a). Comparative studies on Painted Stork nest survival rates in different climatic regimes across India would be helpful in clarifying their effects.

Since some of the colony-specific parameters like nest density appear as significant in our models and positively affect nest survival, we are inclined to view our results as complementing the traditional cost–benefit approach of avian coloniality (Wittenburger and Hunt 1985). However, it is interesting to note that while the strength of the relationship is weaker in the case of nest density, with a very slight decline in DSR at lower densities, other covariates such as nest age and minimum temperature can cause considerably larger variations in DSR (Figures 4.12 and 4.13).

As our study covered two consecutive breeding seasons, one of which experienced a normal monsoon and the other a poor one (low rainfall), an

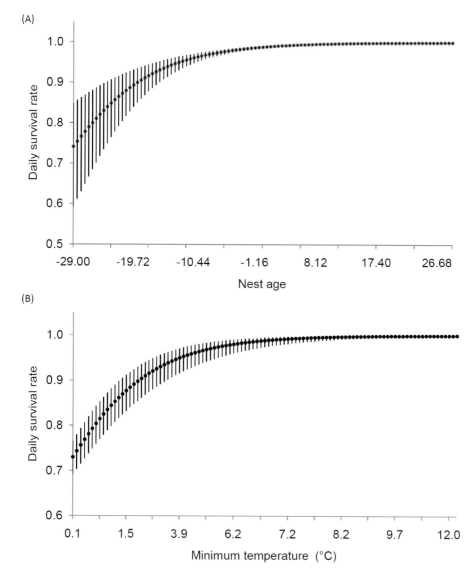

FIGURE 4.12 The effect of (top) nest age and (below) minimum temperature on estimates of daily survival rate (ordinate) of Painted Stork nests. (Source: Tiwary and Urfi 2016a)

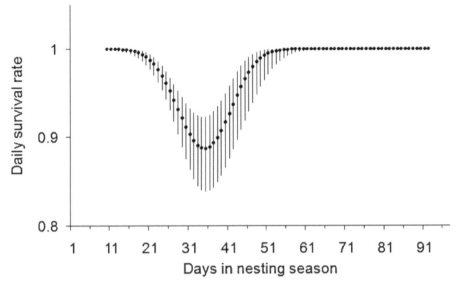

FIGURE 4.13 Variation in daily survival rate of Painted Stork nests across the nesting season. Estimates of daily survival rate (DSR) taken from the model with a quadratic term for DATE where day 1 is the first day in the nesting season. (Source: Tiwary and Urfi 2016a)

opportunity presented itself to study interannual variations in nest survival. Some recent studies have attempted to demonstrate the impact of annual rainfall (Djerdali et al. 2008), and consequently the impacts of local hydrology, on the breeding success of heronry birds (Bino et al. 2014). Clearly, the size of the waterbodies in the vicinity of nesting colonies is important in terms of influencing nesting success. The amount of yearly rainfall creates a strong trophic effect since rains are the primary driver of food cycles, particularly in terms of triggering plankton cycles, as plankton constitute the prey base for fish larvae (Jhingran 1982). The rains also create islands on which colonies can be established, effectively isolated from ground predators, as well as dispersing their food far and wide due to flooding. If the monsoon pattern is influenced, as it is feared to be, by global climate change (Loo et al. 2014), then populations of fish-eating birds like the Painted Stork may be adversely affected.

Chapter 5

Sexual Size Dimorphism and Mating Patterns

Biologists have long puzzled over how sexual size dimorphism (SSD) evolves and whether individuals look for particular attributes, such as a similarity in body size, when selecting a mate. These vexing questions are age-old and among the early thinkers on this subject was Charles Darwin who wrote about body-size differences between the sexes in his book *The Descent of Man and Selection in Relation to Sex*. Interest in the evolution of SSD has not waned and continues to excite evolutionary biologists today (Andersson 1994; Fairbairn 1997; Székely et al. 2007).

In this chapter, we will restrict our discussion to birds and describe a field study in Delhi Zoo which attempted to demonstrate SSD in this Painted Stork population. We also examined the implications of inter-sex differences in different body parts. But first, by way of background, it should be stated that in most species of birds the male has been shown to be larger than the female (Selander 1966, 1972). There are some exceptions to this general rule, notably raptors, in which the female has a larger body size compared with the male, for reasons (related to flight and aerodynamics) that are well understood. While SSD can be primarily attributed to male–male competition, natural selection acting to reduce competition between the sexes (Slatkin 1984; Shine 1989) and fecundity selection (Schulte-Hostedde and Millar 2000) are also proposed as alternate evolutionary hypothesis. Mate preference is cited as an explanation for SSD and it has been speculated that females may prefer larger males primarily because they are more successful in obtaining territories. In territorial disputes the larger male will always have an edge over his rival.

Another question is whether bigger individuals mate with larger counterparts and vice versa. In other words, is assortative mating with respect to body size (or some determinant of body size) the norm? Several studies have reported that assortative mating with respect to body size, body length, and features of wing, bill, tail and tarsus is indeed widespread in birds (Coulter 1986; Chardine and Morris 1989; Sandercock 1998; Wagner 1999; Helfenstein et al. 2004), for which a number of underlying mechanisms have been proposed. These include active mate choice, intra-sexual competition, mating constraints and differential mate availability (Sandercock 1998; Helfenstein et al. 2004). Curiously, in some bird species, mate selection with respect to body size has been reported to be random (Bowman 1987).

Although SSD has been studied in several groups of birds, very few studies have focused on large wading birds like storks and ibis (Kushlan 1977; Bildstein 1987). In the family Ciconiidae the information on SSD is scanty, mostly available through morphometric studies on a handful of museum specimens of which the relevant body size parameters were recorded. From the available information it appears that in many species of stork the male is larger than the female (Hancock et al. 1992) although in most cases the size differences between the sexes are not easily noticeable. With the exception of the Saddlebill *Ephippiorhynchus senegalensis*, in which the female is 10–15% smaller than the male and looks visibly different in the field (del Hoyo et al. 1992), there appear to be few cases of sexual dimorphism among storks. The differences in iris colour in Black-necked Storks – brown/red in the males and conspicuous bright, lemon yellow in females (Figure 1.6) – is a well-known example (Ali and Ripley 1987) of sexual dimorphism unrelated to size or SSD.

During our studies on the Painted Stork population of Delhi Zoo, the question of SSD was considered. Of course, in the field it is impossible to mark out an individual by its sex but observations of mating (copulation) at the nest suggested that the bird on top, likely to be the male bird, appeared larger in size. These differences in body size seemed quite apparent once the birds had disengaged and were standing side by side in the nest, and we began wondering if there was a way to demonstrate SSD in the Painted Stork using a scientific methodology.

A search of the literature revealed that the standard method for studying SSD in birds was to take recourse to dead specimens, dissect them to observe their gonads and thus affirm sex, recording a suite of relevant morpho-metric parameters. (Nowadays, of course, molecular techniques for sexing individuals – said to be quite popular with aviarists – are in vogue, but prior to that the only way to ascertain the sex of an individual was to kill it and cut it open to examine the gonads.) Since the method of analysis usually involved employing multivariate statistical methods, the sample size had to be quite large. At the time when I became interested in this problem, among the studies already published quite a few were on different species of gulls, probably because obtaining dead gull specimens was not difficult given how common these birds are at coastal sites. Large numbers of them that die of natural causes can be easily picked up, which avoids having to obtain a permit for killing living birds for research purposes. All of this is not possible with the Painted Stork: culling large numbers of individuals of a large bird that is on the endangered list for the purpose of a scientific investigation is out of the question (at least in India) and, if undertaken, is likely to invite sanctions (possibly imprisonment) of the strictest kind.

Studying SSD in the Painted Storks of Delhi Zoo

Study design and methodology

The close proximity at which the birds can be seen in Delhi Zoo is extremely convenient, making it an excellent site for conducting a variety of studies using non-invasive methods. A novel approach was employed in which not a single bird was touched, killed or trapped in order to study SSD. Our approach involved videography, which enables morphometric data to be easily recorded. We conducted our studies in 2004 from August to November, the time when Painted Storks had started flocking to the zoo premises at the commencement of the nesting season, just after the monsoon rains, and a repeat study in 2005. Recordings of 53 and 47 different pairs were made in 2004 and 2005, respectively. In 2004, the first and last recorded copulations were on 31 August and 23 November, respectively, and in 2005 they were on 13 August and 6 November, respectively. This gives an idea about the period when Painted Stork are reproductively active at the site.

Images of Painted Storks copulating at the nest could be easily captured by using a video camera mounted on a stand (Figure 5.1). After the storks disengaged, the birds were observed standing side by side and, while we could be certain which was which, further photographs were taken – from which one to three images of each pair were selected for detailed analysis (Rising and Somers 1989). In order to minimise error in measurements resulting from the relative positions of the birds, only those images in which the male and female birds were standing parallel and very close to each other, and approximately perpendicular to the camera, were used. Computer software (see Urfi and Kalam 2006 for further details) was used for quantifying the dimensions of various external body parts from the images. Measurements obtained from video images were calibrated using two stuffed Painted Stork specimens.

FIGURE 5.1 Painted Stork copulating. At Delhi Zoo it is possible to obtain reliable measurements of external characters in a wild population of Painted Storks by using innovative non-invasive field methods. Using videography, at a recording distance of 43–76 m, it was possible to investigate SSD and mating patterns. (Vineet Kumar)

Thickness of the tibia was used as a scale for converting the measurements of the various body parts. Vernier callipers were used to measure the tibia from a point midway along its length. The mean tibia thickness was estimated as 1.44 cm from the two specimens.

A couple of caveats with respect to our methodology are that copulation could not be assumed to be evidence of a successful pair bond, though there was a high chance of it being so. Certainly, extra-pair copulations are suspected to take place (see Chapter 4) and they could also be between birds one of which was below the age of first reproduction (four years). Secondly, since the material used was video images and not live specimens, we did not have the freedom to choose all the relevant morphological features for the study. The actual lengths of different body parts were established using the methodology described by Bosch (1996) and Wagner (1999). The parameters recorded are shown in Figure 5.2 and include the following:

Bill length, estimated distance from the tip of the upper mandible to the corners of the mouth.

Tibia length, estimated distance from the joint of the tibia-tarsus to the feathers.

Tarsus length, estimated as distance between the tibia-tarsus joint and foot. On each bird, separate measurements were made of each leg.

Two additional characters, pertaining to the feathered body parts of the bird, were also estimated. Body length, estimated from a point corresponding to the juncture of the neck and the breast to the approximate base of the tail.

Body depth, estimated from the highest point on the back to the belly.

The Pearson's correlation coefficient between the real and estimated values (of the museum specimens) was high ($P = 0.99$) and the differences between the mean estimated and actual values was not significantly different. However, the video method tended to underestimate the size of the feathered parts by approximately 1.23 cm and so a correction factor was incorporated in the final conversion of the feathered body parts into metric units. For each character, dimorphism index (DI) was calculated as female mean/male mean × 100, in accordance with the methodology described by Greenwood (2003). With the data at hand we could estimate DI for both paired and unpaired sets (Table 5.1).

Broad conclusions of the study

Sexual size dimorphism

The first objective of our study was to confirm something that was already apparent from general observation, which is that in a copulating pair the male tends to be larger than the female. From the data displayed in Table 5.1, we can understand the extent of these differences. Since there was no significant

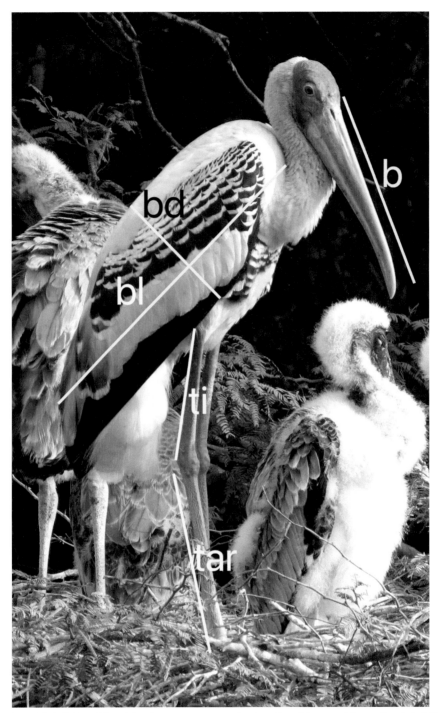

FIGURE 5.2 Painted Stork body parts measured using the videographic method: *b*, bill length; *bl*, body length; *bd*, body depth; *ti*, tibia; and *tar*, tarsus. See text for details. (N. K. Tiwary)

TABLE 5.1 Estimated sizes (cm) of various body parts of male and female Painted Storks studied at Delhi Zoo 2004–2005. Source: Urfi and Kalam (2006)

Character	Sex	N	Mean[a]	SD	CV(%)	DI(%)
Unpaired comparisons						
Bill length	♂	88	26.60	1.09	4.09	88.65
	♀	85	23.58	1.54	6.53	
Tibia	♂	68	20.90	2.06	9.86	91.00
	♀	54	19.02	2.32	12.19	
Tarsus	♂	60	25.51	2.54	9.96	92.90
	♀	41	23.70	2.32	9.79	
Body length	♂	54	52.61	7.02	13.34	90.71
	♀	49	47.72	6.04	12.66	
Body depth	♂	56	27.94	3.88	13.89	90.26
	♀	53	25.22	2.72	10.78	
Paired comparisons						
Bill length	♂	81	26.58	1.11	4.18	88.34
	♀	–	23.48	1.48	6.30	
Tibia	♂	48	20.81	1.98	9.51	90.10
	♀	–	18.75	1.99	10.61	
Tarsus	♂	36	25.96	2.29	8.82	90.64
	♀	–	23.53	1.96	8.33	
Body length	♂	39	52.16	6.70	12.84	90.09
	♀	–	46.99	5.70	12.13	
Body depth	♂	45	28.38	3.92	13.81	88.97
	♀	–	25.25	2.67	10.57	

SD standard deviation; CV coefficient of variation; and DI dimorphism index. The paired and unpaired data sets are shown separately.

[a]ANOVA P < .001

yearly difference in the mean sizes of various body parts, the data for 2004 and 2005 were combined for a detailed statistical analysis. Our data revealed that the variability of sizes in body parts between sexes was greater in the soft body parts than in the hard skeletal parts. Furthermore, the variability in most characters was greater for females.

The highest differences were recorded in the dimensions of bill (DI = 88%). A principal component analysis (PCA) revealed that the first two components account for 85% and 88% of the variation in the male and female, respectively (Urfi and Kalam 2006). The first principal component (PC1), which accounts for most of the variation, can be interpreted as a size axis in which the maximum correlation is observed for body length. Using PC1 scores as an indication of overall body size, we tested for differences between the male and the female. An ANOVA of the two PC1 samples (i.e. for males and females) revealed differences ($F = 4.46$, df = 1, 46, $P < .05$) indicating significant SSD (Figure 5.3). A test of normality revealed that although the PC1 scores of both sexes are normally distributed, the skewedness of the female scores is greater than the male scores.

While the tarsus is a good indicator of body size in many birds (Rising and Somers 1989), differences in other external factors can also be relevant. For instance, in our study, the PCA showed that body length is actually a better indicator of overall body size, compared with other characters, because of its high correlation with PC1 score in both sexes (Figure 5.4). Our observations of the nesting behaviour of Painted Storks in Delhi Zoo confirm that most social interactions are intraspecific in nature and their intensity is highest at the beginning of the nesting season. In territorial fights, larger individuals (males) are likely to have an advantage over smaller birds. It is important to bear in mind that the 9–10% difference in body length between sexes, as shown in our study, translates into a much greater difference in terms of body mass considering the allometric relationship between length and mass in animals. Greater male size is favoured by natural selection up to an optimal body size beyond which greater size would be a disadvantage. However, since an increase in body size could also be a function of age this aspect needs to be

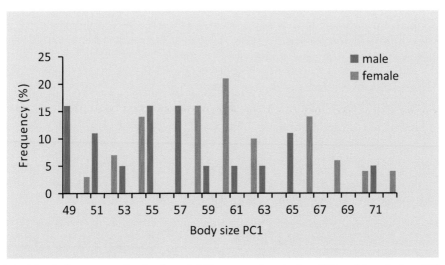

FIGURE 5.3 Frequency distribution of PC1 scores of male (n = 29) and female (n = 19) Painted Storks. PC1 scores were obtained from a PCA of five variables of external body parts. (Redrawn from Urfi and Kalam 2006)

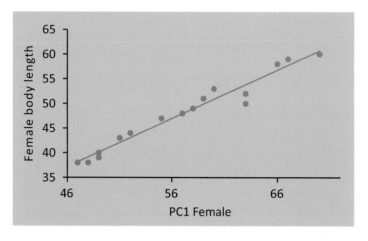

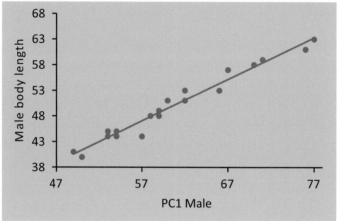

FIGURE 5.4 Body length (cm) as a function of overall body size as determined by a PCA on five variables of external body parts of male and female Painted Storks. (Redrawn from Urfi and Kalam 2006)

further investigated. Studies on the relationship between body size of the male and female birds and their relationship with fitness could also be an interesting line of investigation. As we will see later in this chapter, birds that breed early in the season are larger than those individuals that breed later, with important implications for nesting success.

A trophic organ like the bill can be indicative of selection, although enormous plasticity has been reported in this structure. A high value of DI in the case of bill length is likely to have important consequences with respect to the foraging abilities of the two sexes. In White Ibis breeding colonies, males are more likely to pirate the prey of other birds and are less likely to be victims of piracy than females, presumably because they are better able to intimidate others (Frederick 1985). Larger bill size in males could result in more efficient harvesting of food resources, which in turn could lead to greater fitness (Bildstein 1987). For the Painted Stork (see discussion on prey capture and bill shape in Chapter 6), the primary demand on the trophic organ may be to get a good grip on prey which is slippery and strong and capable of easily escaping after having been caught, and so a larger bill may well be advantageous. A larger bill could also be more efficient for gathering nesting material, such as pulling off the twigs needed for nest construction. Larger tarsus and bill size may be helpful, too, in exploiting deeper foraging habitat. It has been suggested that morphological differences can lead to differential foraging, which in turn helps in reducing inter-sexual competition for food (Selander 1966). To what extent this may be the case in Painted Storks remains to be studied.

Nests continued to be built all through the season, from August to November. Since we had the date for each copulation bout caught on camera, it was possible to plot male body parts against date to test whether the birds that nested early had bigger body sizes compared with those which bred later in the season. Clearly there were many birds that bred early; in fact, right at the beginning of the nesting season there was huge competition for nesting sites and plenty of fights. We estimated that the rate of intra-specific interactions – which were mostly fights between rivals for territories or nesting material – was very high in the month of August and declined in the following months. Our data indicated that birds who were building nests early on were those with larger (estimated) body parts, compared with those which came later.

As a further exercise, we grouped individuals into early and late breeders. Using the 2004 data only (data on all variables was available for a longer period during this year), we classified early nesters as those individuals which were recorded copulating before 1 September. The rest of the data was grouped under late nesters and the two sets of data were used for studying the differences between early and late nesters. With respect to body length, no significant differences were observed for either males or females. However, we found that median tarsus length of males among the early nesters was significantly greater (Mann Whitney test $P = .001$) than those of later copulating male birds. For females the difference between the tarsus length of early-nesting and late-nesting birds was not significantly different. Whether the later birds were individuals which started nesting early, failed and then started another brood or were attempting extra-pair copulations was difficult to ascertain as the birds under study were not individually marked.

Assortative mating

Large birds would be expected to mate with large individuals and vice versa. To study mate-size selection patterns in the Painted Stork we plotted the values of four different body variables for males and females using paired data sets. The details are shown in Figure 5.5 The linear regression equations for each case are as follows:

Female body length = 16.0 + 0.595 male body length, $r^2 = 0.49$, $P < .001$
Female bill length = 0.73 + 0.856 male bill length, $r^2 = 0.41$, $P < .001$
Female tibia length = 8.03 + 0.515 male tibia, $r^2 = 0.26$, $P < .001$
Female tarsus length = 15.3 + 0.315 male tarsus, $r^2 = 0.13$, $P < .05$

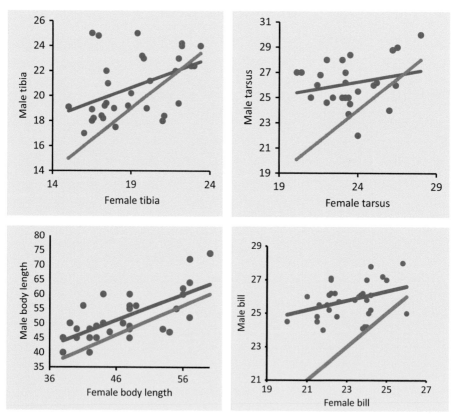

FIGURE 5.5 Relation of lengths (in cm) of tibia, tarsus, body length and bill length of male and female Painted Storks recorded on video during copulation at Delhi Zoo in 2004–2005. The *blue lines* show the trend line (significant at $P < 0.001$ in all cases); *orange lines* show the condition $x = y$ in all graphs. Note that the abscissa and ordinate scales are different in each graph. (Redrawn from Urfi and Kalam 2006)

Our results are suggestive of positive assortative mating. Several factors have been attributed to the evolution of assortative mating, namely active mate choice, intra-sexual competition and mating constraints/differential mate availability (Coulter 1986; Sandercock 1998; Wagner 1999), and this subject needs to be explored in detail in the case of storks. Variability is observed in other characters as well (Figure 5.5) and could be due to biased sex ratios in the Painted Stork population, with there being more females than males. While more studies are required on mating patterns, an important point to be borne in mind as far as this study is concerned is that the data obtained from videography was essentially morphometric recordings of birds sexed during copulation, by their relative positions. It was assumed that the pair engaged in copulation was a pair associated with a particular nest but the possibility of two individuals engaged in an extra-pair copulation also cannot be ruled out – nor trios, as has been reported in related birds (Fujioka 1986; Datta and Pal 1995).

Patterns of nest location

To a casual visitor seeing Painted Storks nesting in trees at Delhi Zoo, it will not be immediately obvious that there are definite patterns in how different tree clumps or nesting substrate are occupied. The zoo has two main ponds (1 and 2) that each have islands on which mesquite trees have been planted. The merged canopies of trees in pond 1 appear to be larger compared with those in pond 2 (see Figure 3.15). It would appear that more surface area is available for Painted Stork to place nests in Pond 1, compared to pond 2. Yet our observations indicate that the nest density of the colony at pond 2 is higher than that of pond 1.

Besides the islands in the two ponds, Painted Storks also build nests on neighbouring trees in the vicinity of ponds. At Delhi Zoo, in each of the years of study we found a characteristic colonisation pattern. In mid-August, at the time when Painted Storks start congregating at the zoo, the first arrivals settle at the two ponds and the majority of the nests are built within a couple of weeks (there is another bout of nest-building in October) (Figure 5.6). In some years colonies at pond 1 are colonised first with the remaining birds colonising pond 2, but this has not been conclusively established. Having said that, we are slowly improving our understanding of the system at Delhi Zoo so that and it can be used as a model for detailed studies in future.

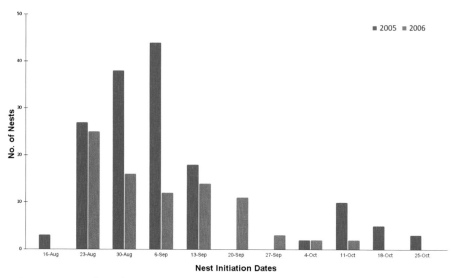

FIGURE 5.6 Number of new nests built by Painted Storks in the Delhi Zoo colonies during 2005 and 2006 in ponds 1 and 2. (Redrawn from Meganathan and Urfi 2009)

BOX 5.1 Studying Painted Stork morphometrics using non-invasive methods: asking biogeographic questions

The growing popularity of employing non-invasive techniques, using digital images to study behaviour and morphometrics, is a testimony to the ease and convenience which such approaches offer. The cumbersome procedures of handling live animals and obtaining permits for their capture are easily avoided. Possibly inspired by our videographic studies on SSD of Painted Stork, Mahendiran et al. (2022) successfully measured the body sizes of wild Painted Storks in two different biogeographic regions of India, spatially separated by more than 2,000 km along a north-south axis in India. The main objective of their study was to explore the relationship between local bioclimatic conditions and the morphological variations in wild Painted Storks from two separate ends of the country. Birds and mammals (endotherms) face different challenges in different environments to maintain a constant body temperature and minimize energetic costs. For instance, Bergmann's and Allen's rules state that the organisms in colder environments and those in higher latitudes and altitudes tend to be larger with smaller appendages to minimize heat loss, compared to those living in warmer areas. Evidence of the operation of eco-geographic rules between the regions rendering the body size variation and sexual dimorphism in the Painted Stork from the two different clusters was conclusively demonstrated in this study.

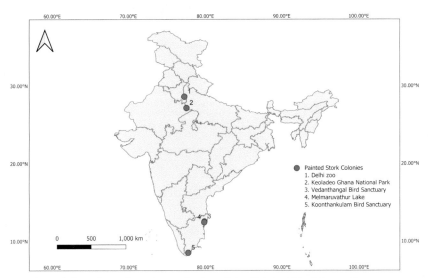

Location of the selected nesting colonies of the Painted Stork in northern India (1 & 2) and South India (3, 4 & 5) from where morphometric measurements of select body parts were recorded photographically. The two clusters of sites are separated by a distance of at least 2,000 km. (Redrawn from Mahendiran et al. 2022)

Chapter 6

Foraging Ecology

A continuous and regular supply of food is of critical importance for birds to breed successfully and raise their young. In the Indian context, the monsoon is a lifeline for all forms of biodiversity and for the country's economy (Box 6.1). These seasonal rains trigger the food cycles of terrestrial birds, via the proliferation of insects and other invertebrates which provide animal protein for the growing young ones, even for those species that are not insectivorous as adults (Padmanabhan and Yom-Tov 2000). The monsoon has a similar triggering effect on food cycles in wetlands. It is necessary to emphasise this point right at the start of this chapter because it helps us to understand the foraging adaptations and behaviour of Painted Storks in an ecological context.

BOX 6.1 The Indian monsoon: a life giver for humans and Painted Storks

Among the most predictable of all storm systems, the monsoon is a shifting seasonal wind pattern that brings months of near-constant rain after several months of clear, hot, dry weather. Perhaps the world's best-known monsoon is the one that blows across the Indian Ocean. With the onset of summer in April the land heats up more rapidly than the ocean, and monsoon winds begin to blow inland across the Indian subcontinent. A low-pressure area forms as heated air above the land expands and rises, and warm, moist ocean air moves in to take its place. Passing over the hills and highlands, the ocean winds then drop their moisture as torrential summer rains known as the south-west or summer monsoon. In autumn, over the southern part of the subcontinent, the situation is reversed as the land cools down more rapidly than the sea. From October to December the monsoon winds blow steadily seaward, bringing rain to some places in South India.

The Indian summer monsoon spans a full three months from June to September. It provides India and its immediate South Asian neighbours with almost 80% of their annual rainfall. A major source of household water and important for hydroelectric power, the rains are vital to the heavily populated, primarily agrarian subcontinent. The monsoon not only sustains plant and animal life in all its diversity but also has a profound effect on national economies and polity in this region. In India, agriculture is the largest and most important sector of the economy. The rainy season is critical to the production of grains for human consumption and for animal fodder, which supports the world's largest population of

The arrival of the Indian monsoon is an important event for both humans and biodiversity. 'After months of searing sun, when all is heat and dust, the arrival of the Indian monsoon is an event. In June, dark, bloated clouds loom overhead, and when they finally release their load of rain, everyone welcomes the sweet, musty smell of wet earth. The brown landscape suddenly turns green, and dried up rivers fill, swell, and overflow their banks. The rains trigger the spawning of fish, amphibians, and other aquatic creatures, which are carried far and wide by the floodwaters. Soon the surplus of edible freshwater species stimulates more new life in the wetlands. Among the fish-eating birds that take advantage of the bounty are painted storks.' Extract from *A monsoon delivers storks* (Urfi 1998). (A. J. Urfi)

dairy cattle. A monsoon failure means devastation for the economy and also for the environment. The erratic behaviour of the monsoon in recent years, characterised by extreme weather patterns – suspected to be due to global climate change – is highly worrying. The trophic interactions in wetlands are strongly influenced by climate change (Harrington et al. 1999). So, for people and for Painted Storks, a good monsoon makes all the difference.

Trophic adaptations in *Mycteria*

For birds the bill is of primary importance in the context of foraging, although this trophic structure may perform other functions too, for example as a tool for nest construction or as a weapon against rivals. An adult Painted Stork's bill is large (> 24 cm length) with a curved tip (Figure 6.1). The bird does not hold its bill completely horizontal to the ground while standing, thus giving an impression that it is strongly decurved. Actually, the bill is straight for most of its length but approximately 18.5 cm from the mouth there is a curvature of approximately 11°, and a gap of 1 mm or more between the mandibles in a locked position. The line along which the mandibles are locked is sinuous,

FIGURE 6.1 The Painted Stork's bill is large with a curvature at the tip and a small gap between the mandibles when they are in a locked position. (N. K. Tiwary)

with an elevation and depression in the distal and proximal portions, respectively. Thus, the intermandibular gap is variable and at some points along the length of the bill the inner surfaces of mandibles are more or less parallel to each other (Urfi 2011a&b).

The decurved bill of Painted Stork is a characteristic of the genus *Mycteria*. All four species are primarily piscivorous, foraging in muddy waters where capturing fish using non-visual cues is necessary. However, all *Mycteria* take other types of prey opportunistically as and when encountered by employing visual cues. For example, the Wood Stork consumes crayfish, amphibians, insects, small snakes and even sometimes baby alligators, and a Painted Stork has been recorded swallowing a snake. In contrast, members of the tribes *Ciconiini* and *Leptoptilini* have straight bills. *Ciconia* species have a characteristic visual foraging method and a catholic diet comprising of vertebrates and invertebrates. Some species, such as the White Stork, are long distance migrants and exhibit considerable variation in their diet between nesting and wintering grounds. *Ephippiorhynchus* storks have straight, thick, slightly upturned bills with which they primarily catch fish by walking in the shallows and stabbing the bill repeatedly into water (Maheswaran and Rahmani 2002). However, the Saddlebill can be observed foraging in muddy waters, catching prey by tactolocation in a manner similar to *Mycteria* (del Hoyo et al. 1992). The Black-necked Stork is also known to employ tactile foraging but the fish it captures are generally brought to dry ground and then swallowed, in contrast to *Mycteria* species, which complete all foraging activities in the water. *Leptoptilos* storks have straight, thick and slightly upturned bills and forage visually on fish and a variety of other items, including offal at rubbish dumps.

In many species foraging is a highly specialised activity and correspondingly there are morphological adaptations in the foraging apparatus (Hulscher 1996). The functional morphology of down-curving bills of waders such as ibises and curlews has been the subject of several investigations, and explanations about their shapes have been proposed on the basis of mechanics (Davidson et al. 1986, Bildstein 1993, Owens 1984). We can only speculate on the how these features of trophic apparatus help the Painted Stork to catch fish. The following explanation for its slightly decurved bill of Painted Stork seems plausible. Compared with a hypothetical straight bill, the points of contact between the mandible and prey are more extensive in a curved bill (Figure. 6.2), resulting in a stronger grip on the prey. The pincer-like tip is probably an adaptation for obtaining purchase and preventing prey from slipping free. Given that fish produce powerful jerks by contractions of their

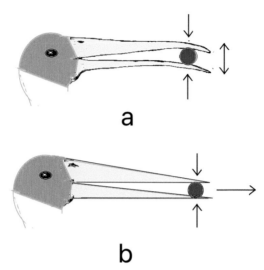

a

b

FIGURE 6.2 Model to explain the functional morphology of a Painted Stork's bill. (Top) Diagram of a Painted Stork holding a prey item at the tip of the bill. As the mandibles press the fish it is flattened and a larger surface comes in contact with the bill. Due to the curved shape of the bill, the mandibles are more or less parallel to each other, thereby permitting greater contact with the prey. (Below) Depicts the situation with a hypothetical straight bill. As the mandibles press against each other the resulting pressure would push the prey forward and it may be lost. A much lesser surface area of the mandibles comes in contact with the prey, so the grip is weaker. (Redrawn from Urfi 2011a and b).

body, this is likely to be an important consideration; possibly due to this adaptation, once captured by the Painted Stork fish do not escape easily unless dropped.

Foraging behaviour

The general pattern of tactile foraging by *Mycteria* species is that the foraging individual inserts its partially open bill into the water, keeps its eyes above the surface and moves forward. Sometimes the bill is moved from side to side, and foot-stirring, as also recorded in egrets, is also employed. Along with wing-flashing, foot-stirring is presumably used to startle concealed fish. As the water in which the birds forage is muddy, the birds probably cannot see their prey clearly (Ogden et al. 1976). The Painted Stork sometimes pivots around a point, in the manner of ibises. This behaviour is recorded more frequently in vegetated habitats but its functional significance is unclear. The typical manner of Painted Stork foraging can be termed 'active tactolocation'; in other words, the foraging bird walks in the water with its bill immersed (Box 6.2).

BOX 6.2 Tactolocation in *Mycteria*

In a study on the reflex action of the mandibles in the Wood Stork, Kahl and Peacock (1963) demonstrated that the lower mandible snaps shut extremely rapidly (approximately 25 milliseconds) as soon as it meets any object. The *Mycteria* bill has some tacto-receptors, whose exact location is not clear. The Wood Stork is able to forage in total darkness using tactolocation and, assuming that the basic mechanisms are the same in all *Mycteria* species, this may explain why the system is so effective in muddy waters and why the Painted Stork should also have no difficulty foraging in darkness. However, while successfully explaining many aspects of tactile foraging in *Mycteria* storks, the bill snap reflex model leaves some important questions unanswered (Kushlan 1978; Coulter and Bryan 1993; Urfi 2011a&b). For example, while foraging in a vegetated area, it would be impossible to avoid the mandibles meeting some object accidentally, yet we do not see the bill needlessly snapping shut. Additionally, when moving through tangled vegetation, the bird does not seem to get confused between fish and plants. Fish over 3 cm are taken in preference; so, clearly, this is a system which is at least partly selective. How this can happen in a system based on tactolocation is unclear.

Glut of fishes after the monsoon (A. J. Urfi)

However, sometimes the bird stands stationary, seemingly waiting for fish to make contact with the mandibles (passive tactolocation).

The Painted Stork, being restricted to the littoral zone, forages in areas which are less than 25 cm deep (Figure 6.3). Typically, their foraging activity is in bouts of varying duration, followed by periods of inactivity when the bird either stands on one leg or hunched up, or sits on bent tarsi on the shore (Figure 6.4). The periods of inactivity could be due to the bird being satiated. Alternatively, these periods could be dictated by factors linked to prey availability and accessibility, such as differences in the activity patterns of fish. However, through the day, the oxygen content of the water fluctuates in relation to surface temperature and this has an effect on fish behaviour. Thus, during early morning, when the dissolved oxygen content of waters is low, fish may come to shallow areas, or close to the surface or make more frequent trips to the surface for gulping air and in the process become vulnerable to capture. At other times of the day, fish may go into deeper areas, out of the reach of storks (Kalam and Urfi 2008).

FIGURE 6.3 The Painted Stork, being restricted to the littoral zone, forages in areas that have a water depth of under 25 cm. (Paritosh Ahmed)

FIGURE 6.4 Painted Storks resting on a bank, either because they are satiated or waiting for a time that is favourable for foraging. Note some birds are sitting on their hocks. (N. K. Tiwary)

Since Painted Storks are colonial nesters they would be expected to be flock foragers, as nesting colonies also serve as information centres about the location of food patches. However, foraging group size is variable. In the nesting season flock sizes are smaller than in the non-breeding season, probably because as the ponds and wetlands dry up in summer birds tend to concentrate on the few remaining ones. Age and season-specific differences in the foraging abilities of different individuals in a population have also been observed. For example, in the non-breeding season juvenile birds have lower, though not significantly different, foraging success compared with adults, which could be due to lack of experience in the young birds. The prey capture rate of adult Painted Storks in the breeding season has been estimated to be nearly two and half times higher than in the non-breeding season, indicating varying energetic demands at different times of the year.

Diet and prey size

Two studies from the Delhi region have reported in detail on the diet of Painted Storks (Table 6.1). In the first, eleven species of fish, some being those of commercial significance, were identified through gut content analysis (Desai et al. 1974). In the other, a video study, Mozambique Tilapia were observed to be part of the diet (Kalam and Urfi 2008), although accurate identification

TABLE 6.1 Fish species recorded in the diet of Painted Storks from two studies in the Delhi region. (Source: Urfi 2011b)

Species	1966–71 Gut content analysis	2004–06 Videography
Wallago *Wallago attu*	+	+
Stinging Catfish *Heteropneustes fossilis*	+	−
Giant River-catfish *Sperata seenghala*	+	−
Scarlet-banded Barb *Puntius amphibius*	+	−
Catla *Catla catla*	+	−
Mrigal Carp *Cirrhinus cirrhosus*	+	−
Rohu Labeo *Labeo rohita*	+	−
Spotted Snakehead *Channa punctata*	+	+
Striped Snakehead *Channa striata*	+	−
Bronze Featherback *Notopterus notopterus*	+	−
Clown Knifefish *Chitala ornata*	+	−
Tire Track Eel *Mastacembelus armatus*	−	+
Mozambique Tilapia *Oreochromis mossambicus*	−	+

of fish captured was not possible in most cases. Tilapia is an exotic fish which was introduced to India in 1952 and has since become a major invasive species (Jhingran 1982). The absence of tilapia in the diet of Painted Stork in the mid-1960s (Desai et al. 1974) and its occurrence in a later study (Kalam and Urfi 2008) could be indicative of fluctuations in its relative abundance in North Indian wetlands over the years. Besides ubiquitous fin-fish, items such as fragments of plant fibre and other plant material, including leaves, as well as small stones, aquatic insects and a frog have also been reported in gut content studies on the Painted Stork (Desai et al. 1974). Since storks are almost all exclusively carnivorous, plant material in their gut is most likely taken incidentally while foraging and is unlikely to be of any dietary significance. At Kokkare Bellur, adults were observed feeding nestlings with frogs, crabs, large insects and grasshoppers. While Black-necked Stork and Greater Adjutant Stork have been recorded catching birds, nothing so spectacular has been reported for the Painted Stork. At coastal sites in India, Painted Storks have been observed foraging on crustaceans and invertebrates common to such waters. I have myself observed large numbers of Painted Storks feeding in salt pans and mudflats on the coast of the Gulf of Kutch in Gujarat. Visual foraging should be expected more often in such situations.

The relationship between prey size and handling time (Krebs and Davies 1984; Stephens and Krebs 1986) was investigated in the Painted Stork using video techniques (Kalam and Urfi 2008). Since small prey is easily swallowed, negligible time is spent in handling it, but with increasing prey size, handling time increases exponentially (Figure 6.5 top). Thus, beyond a point, large prey become unprofitable, even though they may have a high calorific value (Figure 6.5 below). Different species of fish have different calorific values and so their nutritional content may vary considerably, but fish larger than 20 cm were rarely captured and most of the prey ranged between 1 and 12 cm. There are recorded instances of Painted Storks dropping very large fish after

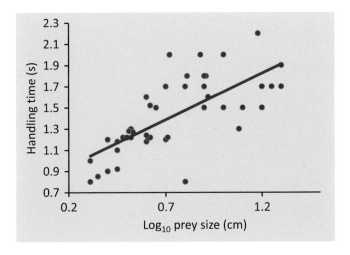

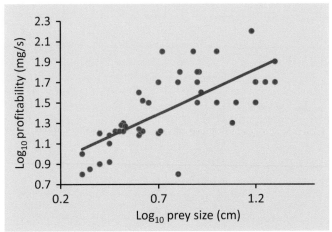

FIGURE 6.5 (Top) Relationship between dry weight and handling time of prey taken by Painted Storks at different foraging sites in 2004–2006. (Below) Relationship between length and profitability (prey dry weight/handling time) of prey taken by Painted Stork. The equation for estimating dry weight from prey length was derived from a length–weight relationship of the fish species Spotted Snakehead *Channa punctata*. (Redrawn from Kalam and Urfi 2008)

capturing them. While this could be due to energetic factors (low profitability), the possibility that large fish can have sharp structures on their fins, which may injure the predator, could also be an explanation.

Differences in size of prey captured during the breeding and non-breeding seasons have also been recorded (Figures 6.6 and 6.7). However, whether such differences are due to active selection or are of more a reflection of

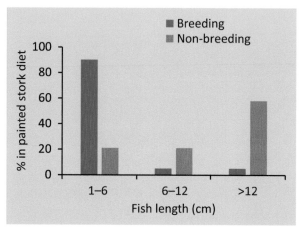

FIGURE 6.6 Length of fish taken by Painted Storks at different foraging sites in breeding and non-breeding seasons in Delhi in 2004–2006. (Redrawn from Kalam and Urfi 2008)

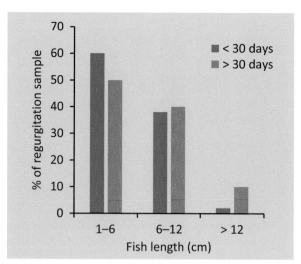

FIGURE 6.7 Length of fish regurgitated to younger (< 30 days) and older (> 30 days) chicks by adult Painted Storks at Delhi Zoo in 2004–2006. (Redrawn from Kalam and Urfi 2008)

seasonal changes in the abundance of certain size groups of prey remains to be established. In the monsoon months, due to spawning activity, small fish predominate, but in the summer months larger fish are abundant. The size of prey regurgitated by Painted Storks during the period when the nestlings were very small was significantly smaller than those provided a few weeks later when the nestlings had grown. If a fish larger than 20 cm in length was regurgitated in the nest it was ignored by the nestlings, suggesting a gape size constraint. So large fishes are the exception, and usually taken by individuals who are not catching food for their chicks (Figure 6.8). Having said this, a certain amount of predigestion happens in the crop, so even large prey can be broken down into smaller pieces.

Given that field observations of birds are generally made during the day, instances of nocturnal foraging usually go unrecorded. However, some instances of night foraging from different parts of India are on record. Next to actively foraging, an alternative strategy common among carnivorous birds like storks and herons is to steal food (Frederick 1985). While several bird species occur alongside Painted Storks at foraging grounds, among fish-eating waders of a comparable body size, Grey Heron and Great Egret are the only birds from which a Painted Stork can steal prey. Mahendiran and Urfi (2010) observed Painted Storks stealing fish from Little Cormorants. However, the fish taken

FIGURE 6.8 Large fish prey, as in this picture, are the exception rather than the norm. In the nesting season the bulk of the fish caught are smaller. (N. K. Tiwary)

by the cormorants were relatively small since this species forages in shallow waters, often near the shore. The Painted Stork is able to steal prey from other waders, except for Black-necked Storks, but at the nesting site it is unable to defend regurgitated fish from herons and egrets.

Comparisons with the Milky Stork

In South-East Asia, the Painted Stork is largely recorded as frequenting freshwater habitats while the Milky Stork prefers the coast, so the two species are ecologically separate (Li et al. 2006). (This is not to say that the Painted Stork is not found on the coast. The author has personally seen Painted Stork foraging in large numbers on the coast of Gujarat (western India), in the salt pans between Mithapur and Dwarka, along with flamingo and other species.) Furthermore, the Milky Stork is perhaps more likely to engage in nocturnal foraging (Swennen and Marteijn 1987; Hancock et al. 1992). A study from Sumatra reported Milkfish *Chanos chanos*, Elongate Mudskipper *Pseudapocryptes elongatus*, Giant Mudskipper *Periophthalmodon schlosseri* and mullets (Mugilidae) in the diet of Milky Storks (Indrawan et al. 1993; Iqbal et al. 2009). All of these prey being coastal or brackish water fish, their availability in the intertidal zone is largely governed by the tidal cycle rather than the diurnal cycle, so that may account for the Milky Stork being active at night. The Milky Stork, like the Painted Stork, also employs tactile foraging, which involves probing in the mud with a partly open bill, then drawing it in an arc from side to side, as well as foot-stirring. Foraging individuals are usually spaced 50–100 m apart although tight flocks have also been observed. While searching for mudskippers, Milky Storks probe in holes, sometimes immersing the whole bill and head in the mud. Once the bill is inside the hole it is pushed forward, opening up a groove in the mud; any prey which comes into contact is immediately captured as this method is entirely tactile. Such specialised foraging behaviour has not been recorded for Painted Stork.

Nesting time in relation to food availability

From a physiological point of view, three phases are noteworthy in the annual calendar of birds: breeding, moulting and migrating. Since all of these are energetically demanding processes and temporarily reduce the physical activity of birds, they obviously cannot occur simultaneously and one must succeed the other with no overlaps. Generally speaking, all of these phases are timed with the seasons, which have a strong impact on several crucial 'ultimate factors', which influence timing of reproduction. Since food availability is the single most important factor governing nestling survival, the factors that govern the food cycles are extremely important. Hence, the timing of these factors ultimately leads to the evolution of well-defined breeding seasons for different species of birds.

The 'food availability–nesting time hypothesis' – a well-known concept in ornithology – suggests that nesting seasons of birds have evolved to coincide with periods of maximum food availability in the environment. An assured supply of food in abundant quantities is crucial for reproduction in both birds and mammals. For birds the food supply gurantee has to come from the environment but less so in the case of mammals which feed their young on mother's milk. We have some idea about the quantities of food required by Painted Storks during the nesting period from two (albeit very rough) estimates on record. According to Desai et al. (1974), taking basal metabolic rate into account, it was estimated that a colony of about 100 nests of Painted Stork would consume about 24 tonnes of fish during one breeding season (approximately 150 days). This estimate was derived largely from studies undertaken at Delhi Zoo and the basic unit was a family consisting of 2 adults and 1.1 chicks. In the second estimate, for Keoladeo National Park, using a different set of parameters (2,000 nests, family size equal to 2 adults and 2 chicks and a 90-day period), Ali (1953) came up with an estimate of approximately 90 tonnes or 90,000 kg. In this study, since 2,000 nests were only a small fraction of the colonies at Bharatpur, the fish harvest figures for the entire park would be much greater.

Though this information is very old and crudely estimated, it gives us some idea about the quantity of food resources required to sustain a colony in an ecosystem context. Both of the above estimates pertain only to the Painted Stork, and if other species of piscivorous birds are included then the total harvested would be much greater. Moreover, these estimates pertain only to the nesting season, 90 and 150 days. If a whole year were taken into account, then the amount of biomass (fish) removed by predation would be many times higher. Naturally, most of the fish taken during the breeding season would be small in size (1–8 cm), though in the nonbreeding season Painted Storks would be expected to take larger fish. What consequences this has on fish population structure and indirectly on aquatic communities is not known but warrants a detailed study. Given the sheer quantity of biomass that is consumed, the chances of toxins being present in the food chain, particularly pesticides, and their biomagnification, are enormous. Therefore, Painted Storks, being placed at the highest trophic level in their ecosystem, are likely to be affected by contaminated food to a very large extent.

In *Mycteria* species, the tactile mode of foraging appears to be well suited to this situation because it works best if the density of prey, and correspondingly the encounter rate, is high. High prey densities can result either from concentration or the production of fish. In the case of the Wood Stork, essentially a dry season nester, this effect results when waterbodies evaporate and fish become concentrated. Densities of fish in drying pools in Florida have been recorded to be as high as 8,000 fish/m^2 (del Hoyo et al. 1992). In contrast, in the Indian subcontinent, a similar effect results from the glut of fish due to spawning activity after the monsoon. In one study at Keoladeo Ghana National Park in northern India, nearly 336 fish fry were recorded in a plankton net of 1 m^2, operated for 85 seconds (Ali and Vijayan 1983). Thus, an estimated 8,600,000 fish fry were entering the park in one day (Box 6.3). However, at some sites, notably Kokkare Bellur in southern India, Painted Storks are believed to nest in the dry season, although it is not clear to what extent they take advantage of fish which become concentrated as waterbodies dry up, if such a strategy is adopted at all. There seems to be no detailed study on the foraging ecology of Painted Stork from South India.

Painted Storks in an ecosystem context

What impact aquatic birds have on the ecosystem is a question that has only recently come to prominence (Green and Elmberg 2013; Hanson and Kerekes 2006; Kerekes and Pollard 1994; Žydelis and Kontautas 2008). About the role of birds in general, Wiens (1973) commented, 'it seems unlikely that birds exert any major influence on ecosystem structure, functional properties, or dynamics through their direct effects on either the flux or storage of energy or nutrients … their role instead might be as governors or controllers'. So the conventional wisdom has been that birds have some role to play, albeit limited, in the context of either removal of biomass (harvesting), enrichment of the aquatic systems by their droppings and guano, or as dispersal agents for seeds and spores of aquatic vegetation.

The foraging activity of Painted Storks naturally results in the removal of certain amount of biomass from the surrounding ecosystem, and some idea about the quantities of fish harvested by this species, especially in the nesting season, has been provided above. One of the significant impacts of colonially nesting waterbirds is to enrich waters by their droppings. An early study in this regard estimated that the role of aquatic birds, chiefly cormorants, pelicans and other seabirds, in cycling phosphorous in a global context was quite significant (Hutchinson 1950). While foraging Painted Storks are likely to

BOX 6.3 Variations in Painted Stork nesting times across India: relationship with food availability

As the map below shows, there are clearly two clusters of breeding Painted Storks in India. One cluster includes colonies in north and western India. In all of these the nesting is timed with the summer monsoon and commences in August/September and continues until January or February. The other prominent cluster of nesting colonies is in South India. In most of these the nesting period is highly variable and in some cases it seems to be timed with the onset of the winter (withdrawal or north-east) monsoon (Urfi 1998, 2011a). Details of all these nesting sites and some further information are provided in Appendices I and II.

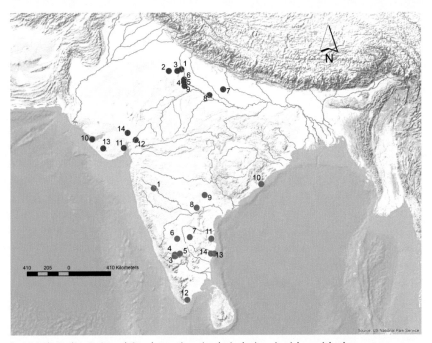

In North India, Painted Stork nesting (red circles) coincides with the summer monsoon while in most of South India (blue circles) it corresponds with either the winter monsoon or other seasonal rains.

enrich the water by their droppings, as individuals are spaced out, the effects of their droppings are likely to be close to insignificant. However, at the nesting colonies, the droppings make an impact due to the sheer numbers of nesting birds involved. Furthermore, in mixed-species heronries, there would be not just the Painted Stork but several other species as well. The only study of this on record seems to be one conducted at Vedanthangal Bird Sanctuary in Tamil Nadu in the 1980s (Paulraj and Kondas 1987; Paulraj 1988). A chemical analysis of the reservoir water indicated that it had high quantities of Nitrate (NO_2) and Phosphorus (P_2O_5) (Paulraj and Kondas 1987).

Many large waders such as ibises and Wood Storks tend to disturb the substrate while foraging. This process, known as 'bioturbation', mixes up the constituents (Bildstein 1993). While Painted Storks and other waders are foraging in the water, it is likely that spores, microorganisms, plankton and seeds of plants will get stuck on their body feathers, feet or even bill and be transported to other sites. Although no study has been done as such, it may be useful to explore what role birds like the Painted Stork have in this context.

BOX 6.4 Food cycles of wetlands

Plankton constitutes the base of aquatic food chains, which go all the way up the trophic ladder to fish-eating birds like the Painted Stork. Long before I embarked on the stork's trail, as a student of zoology I became interested in these tiny, mostly microscopic organisms. I spent hours and hours sitting with a microscope, looking at what lay in a drop of water collected from a drain or a pond. It turned out that a drop of pond water, smeared on a glass slide, was an organic soup full of all manner of interesting creatures. At lower magnifications one could see worms (usually freshwater nematodes), small aquatic insects and other creatures. As I increased the magnification, I could view a variety of gelatinous creatures moving around in the water. These were the plankton – free-floating organisms, with no powers of locomotion of their own. There were some which were disc-shaped with serrations, patterns or markings on their surface and were usually green in colour. Known as 'diatoms', these microscopic plants (which contain chlorophyll) in the group 'phytoplankton' are the primary producers in aquatic ecosystems. Swimming alongside them were tiny animals, 'zooplankton'. These were mostly crustaceans, with calcareous shells or coverings, and included forms such as *Daphnia*, *Cyclops* and others but also rotifers, a fascinating group that includes many forms, such as the genus *Brachionus*.

The presence of live animal food is vital for the growth of fish, particularly in the larval stages when demands for protein are high. Fish, like birds, time their reproduction with the glut of food in the monsoon season. They spawn in large numbers and their larvae feed on zooplankton. For fish-eating birds such as various species of stork, cormorant, heron and egrets – many of which nest in colonies, the basic processes which operate can be summarised as follows:

Monsoon rainfall

↓

plankton bloom and fish spawn

↓

increased fish production (and also dispersal of fish and creation of islands due to flooding)

↓

bird reproduction

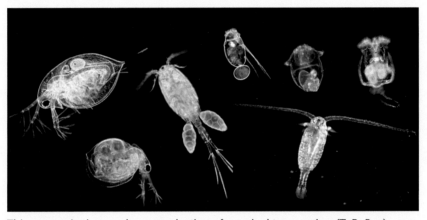

This composite image shows a selection of zooplankton species. (T. R. Rao)

Chapter 7

Painted Storks in an Urban Context

The exponential, often unplanned growth of urbanisation, encroachment on natural habitats and increase in atmospheric and water pollution is causing a host of problems not just for human inhabitants but also biodiversity (Czech and Krausman 1997; Niemela 1999; Savard et al. 2000; Melliger et al. 2018). However, there are also instances of various forms of biodiversity finding habitats in urban sites and some species adapting and even thriving. Some recent studies on common birds of Delhi have shown that certain species of terrestrial and aquatic birds are actually doing well (Tiwary and Urfi 2016b; Rehman et al. 2021), while others are being pushed to marginal areas or driven to extinction.

One of the most important reasons for the decline of many aquatic species of flora and fauna is the disappearance of wetlands, which are regarded as unique repositories of biodiversity and among the most productive ecosystems, crucial for many species of waterbirds (see Box 7.1). Many urban waterbirds are intimately tied to local wetlands for foraging (McClure et al. 2015; Urfi 2006a, 2010b) and breeding (Zipkin et al. 2009). Common birds are effective and extremely popular indicators for evaluating impacts of urban processes on biodiversity because they are diurnally active, easy to monitor and exhibit significant responses to habitat alterations (Bibby et al. 1992; Chace and Walsh 2006; Marzluff et al. 2001; Yang et al. 2015).

Delhi, the capital city of India, has undergone rapid urbanisation over the past several decades due to which human health, safety and biodiversity have been negatively impacted. With the rapid and often unplanned (Taubenböck et al. 2009) growth of urban centres across India, parks, gardens, tree-lined avenues and residential gardens are the only remnant habitats left for birds (Belaire et al. 2015). Much of the metropolis today is a mosaic of built-up

BOX 7.1 Wetlands for survival

Wetlands are repositories of distinct biodiversity which is well adapted to their habitat at the interface of aquatic and terrestrial environments. The Ramsar Convention defines wetland as: 'Areas of marsh, fen, peatland or water, whether natural or artificial, permanent or temporary, with water that is static or flowing, fresh, brackish or salt, including areas of marine water the depth of which at low tide does not exceed six metres.' Note that this definition touches upon all aspects of wetlands and covers a variety of waterlogged areas and other land that is wet. Besides being shallow there is also an element of 'impermanence' of water in a wetland, in that the water level fluctuations can vary both seasonally (say before and after the monsoon rains) and diurnally (due to the tides in coastal areas). So there may be a submerged phase when a wetland does not look like a wetland and a phase at low tide, just a couple of hours later, when the shore is exposed and all manner of living organisms, particularly migratory waders, are attracted to it. The Ramsar definition is quite clear on the 'land' part of term 'wetland', which refers to substrate that can be sand, coral, rocks and so on. The biota of wetlands is quite diverse as animals and plants from both terrestrial and aquatic worlds inhabit them, due to an edge-effect. Fish, inhabitants of the watery world, have several adaptations for an aquatic existence, but typical wetland fish, such as mudskipper can easily live in a moist terrestrial

Wetlands are crucial for the survival of many specialised organisms.
(N. K. Tiwary)

A marsh in North India (A. J. Urfi)

The vast wetlands of the Florida Everglades, USA. (A. J. Urfi)

environment of the inter-tidal zone for limited periods of time (Urfi 2016). Plants, too, possess interesting adaptations, a good example of which is the mangrove. Birds belonging to the orders Charadriiformes (waders) and Anatidae (waterfowl) are generally important components of wetland biota. There are special clauses in the Ramsar Convention which deal with all manner of biodiversity, especially migratory waders and waterfowl.

areas, lined with trees and interspersed with parks and gardens, planted with a variety of indigenous and ornamental flora in an uneven manner. Whereas in some areas (mostly commercial premises as well as some residential localities) buildings and other development dominate, in others there is a mixture of vegetated patches, buildings and roads. Extensive and homogenous patches of native habitat are few in Delhi. The city itself can be said to be largely divided in terms of civic amenities and facilities and the economic profile of residents. All these factors contribute towards the distribution of remaining habitats.

Colonial waterbirds are regarded as bio-indicators of environmental contamination (Fox and Weseloh 1987). Heronry birds that form nesting colonies in cities, of which Painted Stork of Delhi Zoo is a good example, are an important group to study in context of urbanisation. However, the phenomenon of heronry birds nesting in Indian cities is quite common as several other cases are known, of which some examples are given in Table 7.1.

Delhi Zoo: a case study

Although the basic features of Delhi Zoo and its significance as a nesting ground for Painted Storks and other species of heronry birds have been described in Chapter 3, it is worth emphasising that these birds are opportunistic nesters and their regular use of the zoo could be indicative of, among other factors, the shortfall of natural nesting habitats in the wider countryside as well as there being a safe environment within the confines of the site.

The River Yamuna, which lies about a kilometre away from Delhi Zoo, is an important feeding ground for Painted Storks and other birds nesting in the zoo. The stretch of the Yamuna passing through Delhi, though it is only 22 km long and constitutes only 2% of its area, contributes about 80% of the river's total pollution load (Gopal and Sah 1993; Rawat et al. 2003). This is because at least 16 major drains discharge untreated municipal sewage daily into the Delhi segment. Several studies have documented high concentrations

TABLE 7.1 Different species of colonial waterbirds recorded nesting at some Indian cities.

Bird species	National Zoological Park, Delhi[1]	Piele Gardens, Bhavnagar[2]	Kukkarahalli Tank, Mysuru[3]	Karanji Tank, Mysuru[3]	Kankaria Lake, Ahmedabad[5]
Oriental Darter *Anhinga melanogaster*	(−)	−	+	+	(+)
Little Cormorant *Microcarbo niger*	+	+	+	+	(+)
Indian Cormorant *Phalacrocorax fuscicollis*	+	−	+	+	(+)
Great Cormorant *Phalacrocorax carbo*	(−)	−	+	+	(+)
Little Egret *Egretta garzetta*	+	−	(+)	(+)	(+)
Western Reef Heron *Egretta gularis*	(−)	+	(−)	(−)	(−)
Great Egret *Ardea alba*	(−)	+	(+)	(+)	(+)
Intermediate Egret *Ardea intermedia*	+	−	(+)	(+)	(+)
Cattle Egret *Bubulcus ibis*	+	+	(+)	(+)	(+)
Indian Pond Heron *Ardeola grayii*	+	+	(+)	(+)	(+)
Black-crowned Night Heron *Nycticorax nycticorax*	+	+	(+)	(+)	(+)
Black-headed Ibis *Threskiornis melanocephalus*	+	+	+	+	+

Species					
Eurasian Spoonbill *Platalea leucorodia*	+	–	–	+	(–)
Spot-billed Pelican *Pelecanus philippensis*	–	+	+	–	(–)
Painted Stork *Mycteria leucocephala*	+	+	+	+	+
Rosy Pelican* *Pelecanus onocrotalus*	+	–	–	–	+
Asian Openbill *Anastomus oscitans*	+	(+)	(+)	+	(–)

Note: + recorded, (+) presumed to be present though not mentioned in the reference cited, (–) known to be absent from the site, – presence/absence data not available in the source cited, *recorded instances under special circumstances (see Urfi 1997).

[1] Urfi 1997.
[2] Parasharya and Naik 1990.
[3] Rahmani et al. 2016.
[4] Urfi (pers. obs. c.1996) includes birds nesting on trees in Kankaria Zoo (Ahmedabad Zoo).

of heavy metals and pesticide residues (for details, see Urfi 2006a). Dead birds and fish have been reported from this stretch of the river. Furthermore, since this 2010s encroachments on the Yamuna have increased exponentially, as new bridges and flyovers have been constructed. For these reasons both the nesting and foraging grounds of Painted Storks are becoming increasingly isolated (Figure 7.1) and there are fears (though as yet unsubstantiated) that these changes are already impacting the birds.

Our own studies have demonstrated that at downstream sites, such as Okhla Barrage, an important feeding ground for the Painted Stork, located about 9 km from the zoo, there have been marked reductions in the diversity of waterfowl (Urfi 2006a). Indeed, indicator species such as the Sarus Crane *Grus antigone* have not been sighted at Okhla since 1992. Given that there are so many changes happening in their habitat, it is logical to ask how the Painted Storks nesting in the premises of Delhi Zoo are being affected. The data on numbers of storks flocking to the zoo does not indicate a declining trend. In fact there seems to be a slight increase in total numbers recorded, from 438 in 1966 (Desai 1971) to over 500 in 2004 (Urfi 2006a). Whether this is due to traditional nesting colonies outside the zoo, in the countryside around Delhi, having come under the axe, leaving Painted Storks with little option but to head to Delhi Zoo is difficult to say. Further studies on fitness

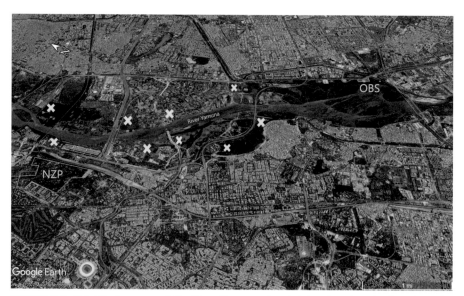

FIGURE 7.1 Satellite imagery of the River Yamuna floodplain, showing the location of Delhi Zoo (NZP) in relation to the river and Okhla Barrage Bird Sanctuary (OBS), the main foraging zone, which are all surrounded by built-up areas. Asterisks mark new development over the past few years. (Google Earth 2023)

parameters such as nesting success will be useful in this regard. The recommendations made by our study group are the following:

• Though there is some degree of awareness about the significance of river floodplain ecosystems (Figure 7.2), a greater integration of ecological concerns in city development plans is required, keeping in view the dependence of birds nesting in the zoo and other sites close to the River Yamuna.

FIGURE 7.2 There is growing awareness about the significance of the river floodplain ecosystem but more needs to be done to conserve it and protect it from encroachment. (A. J. Urfi)

- Unfortunately, there seems to be no official recognition of the value of zoo premises as refugia for endangered wildlife (as well as repositories of architectural heritage) in policy documents (Sharma et al. 2017a; Central Zoo Authority 2023). Clearly there is a need to recognise the significance of these assets and also devise programmes to monitor biodiversity (especially birds) for the benefit of the larger environmental issues they help us to address.

- There is a paucity of education material about the wild waterbirds of the zoo. Facilities for the interpretation of the Painted Stork colonies are virtually non-existent. The few waymarkers and signboards that do exist are outdated or in need of repair. These should be developed in order to achieve the objectives of nature education and a proper research wing should be established. Nesting heronry birds provide a splendid opportunity to teach biology because they offer a view of birds at the crucial breeding stage.

- The Delhi Zoo Painted Stork population probably affords the best example of a detailed study on a single population of a wild bird from anywhere in India (Urfi 2010b). It is recommended that a programme to monitor this population should be initiated and built into the operational machinery of the zoo. Under the guidance of researchers and scientists from academic institutions, this programme could derive its inputs from the efforts of volunteers such as local birdwatchers and school and college students. This would set the tone for generating long-term data sets for breeding birds and simultaneously take a lead in initiation of citizen science programmes for the benefit of the environment (Urfi 2004; Urfi et al. 2005; Greenwood 2007). Since the birds in question are fairly large and their nests easily approached, they are easy to monitor and no special equipment is required to study them. The scientific advantages of carrying out a long-term monitoring programme at Delhi Zoo would be several. It would be useful in understanding population ecology issues, such as how various biotic and abiotic factors influence bird populations, particularly in view of concern about global climate change affecting the monsoon. This data will help us better understand how yearly rainfall patterns affect Painted Stork nesting. Above all, such a monitoring programme would help to study urbanisation and pollution and their possible impacts on birds. Some comparable prominent international programmes are already underway in other countries (see Chapter 9), and a good template for initiating similar volunteer-based programmes in India is already in existence.

BOX 7.2 Adaptations to the urban environment

Birds use urban sites opportunistically and often take advantage of urban structures and features. An occupancy modelling study of common terrestrial birds in Delhi (Tiwary and Urfi 2016b) and along the River Yamuna in Delhi

A defunct thermal power plant a few kilometres from Delhi Zoo, on the banks of the River Yamuna, where Painted Stork used to aggregate around the heated air column when the plant was functional. (A. J. Urfi)

(Rehman et al. 2021) showed that certain species of birds were more likely to be prevalent in certain parts of the city/river and their presence could be correlated with food and other factors necessary for survival found in those areas. With respect to terrestrial birds in those parts of the city that had a high degree of horizontal development, such as old-style bungalows with large gardens and roads lined with fruit-bearing trees (particularly *Ficus* species), woodpeckers, barbets and hornbills were common, but in areas that were built up and had a high density of vertical structures and an absence of green spaces, the bird species composition was markedly different.

Many large-winged birds, such as vultures, storks and cranes, utilise heated air columns or thermals for gaining altitude and then gliding towards their destination with minimal expenditure of energy. In recent years, the exploitation of anthropogenic thermals by large-winged birds for soaring has been noted. A field study by Mandel and Bildstein (2007) on Turkey Vultures *Cathartes aura* feeding at a landfill in eastern Pennsylvania, USA showed that the birds returned to their roosts each evening by gaining altitude in thermals above flared methane vents at the site. Vultures, like storks, are energy minimisers which usually soar and glide when flying between roosts and nesting colonies, to previously located food sources, when searching for new sources of food and during migration (Pennycuick 1972). These birds do engage in flapping flight also but usually it is intermittent, and mostly used when departing from and descending into their roosts or colonies

As the Delhi Zoo colony has been in existence for a long time, it would be expected that the behaviour of birds has changed and they have adapted to the city environment. The zoo is located only a few kilometres away from a thermal power station (now closed). Anecdotal evidence suggests that Painted Storks used to congregate and circle around on the heated column of air that rose above its chimney when the plant was functional. However, the phenomenon could not be scientifically studied before the plant was shut down.

Chapter 8

Painted Storks and People

Storks are the subject of myths and stories the world over, and this is no less the case in India (Dave 1985). In this chapter we will focus on the Painted Stork and their nesting colonies in the context of local communities and whether local villagers perceive them positively or negatively. After all, as concentrations of nests, heronries are often vulnerable to exploitation. For one, they provide an irresistible chance to obtain meat and eggs, necessary for survival in some indigenous communities. But the fishy odours and noise (of chicks) emanating from them also has the potential to make them repugnant. Therefore, it is not surprising that heronries have often been written about in somewhat disparaging terms. They were once referred to as rookeries in early ornithological literature – a reference to noisy, boisterous colonies of rooks. 'Rookery' was in fact a slang term for urban slums in Victorian England (see Box 8.1).

BOX 8.1 Rookeries and heronries in ornithological literature

In literature heronries have been referred to in somewhat disparaging terms. Terms such as heronry and rookery are loaded with historical baggage. The word 'rookery' was probably first used in 1704. According to the Merriam-Webster Dictionary it was a synonym for 'slum' which were described as, 'a thickly populated section especially of a city marked by crowding, dirty run-down housing, and generally poor living conditions'. During this period, when Britain was undergoing industrialisation and people from the countryside were flocking to cities for work, unplanned areas or slums were developing in the cities. By Victorian times slums or rookeries had become the most disreputable and notorious parts of many British cities, and they even find a mention in the writings of the great English novelist Charles

Dickens. For instance, describing the famous St Giles Rookery in London in *Sketches by Boz*, Dickens wrote:

> Wretched houses with broken windows patched with rags and paper: every room let out to a different family, and in many instances to two or even three ... filth everywhere – a gutter before the houses and a drain behind – clothes drying and slops emptying, from the windows; girls of fourteen or fifteen, with matted hair, walking about barefoot, and in white great-coats, almost their only covering; boys of all ages, in coats of all sizes and no coats at all; men and women, in every variety of scanty and dirty apparel, lounging, scolding, drinking, smoking, squabbling, fighting, and swearing.

The Rook *Corvus frugilegus* is a noisy and sociable bird and so it is perhaps understandable that Victorian-age writers and observers found the word rookery – a colony of treetop nests, with loud, garrulous, squabbling owners – the nearest equivalent for a slum. Soon this term came to be used for describing other groupings of animals. For instance, a colony of herons was referred to as a 'heron rookery', though later the term rookery was dropped and it became simply heronry. The term rookery has been widely used in old biological literature to describe colonies of animals as diverse as birds (crows, rooks, seabirds), marine mammals (sea lions) and even some turtles.

Interestingly, heronries themselves, much like rookeries, are not very pleasant places and so they lend themselves easily to negative images, especially when there are chicks in the nests, constantly making irritating noises, begging their parents for food. Their droppings give the place a distinctly fishy odour, contributing to their repulsive character. There are reports from India of villagers cutting down trees on which waterbirds have formed nesting colonies.

Heronry is a broad term used for nesting colonies of storks, ibises, herons, cormorants, pelicans and similar species. (Though for a colony of pelican nests the term 'pelicanry' has also been used.) Not surprisingly, heronries in British India were written about in somewhat disparaging terms by ornithologists. For instance, Jerdon (1864), referred to a pelican colony in South India as consisting of 'rude nests'. Ali and Ripley (1987), who more or less followed the British style of describing the habits and behaviour of birds in their treatise, *Handbook of the Birds of India*, describe waterbird nesting colonies as, 'often twenty nests or more on a single tree crowded cheek by jowl in disorderly tiers'. These authors also employed typically Indian terms in describing specific features of waterbird colonies. For example, with respect to a mixed heronry in North India in which White Ibis tend to form separate conclaves, Ali and Ripley noted that they have 'a tendency to segregation [*sic*] into discrete *mohallas*.' Here, though the context is right in that the ibis does segregate itself and forms sub-colonies within a colony, the term 'mohalla' is perhaps indicative of a stream of thought that subconsciously points toward preconceived notions and stereotypes about specific human communities. Mohallas are neighbourhoods or localities in cities and towns in central and south Asia and often have negative derogatory images associated with them.

'Poverty is the worst polluter' – unfortunately statements like these seem to suggest that poor, marginalized people are the cause of environmental degradation and loss. The formation of slums (as shown in this picture) is a result of a host of factors, primarily due to societal changes. At the time of the Industrial Revolution in Britain, when slums came into existence, their impressions featured in the works of not just writers but also influenced how natural history writers looked at bird nesting colonies. However, just like certain seabirds which nest on islands when suitable substrates are not to be found (passive coloniality), sometimes humans also don't have a choice about where to live. (A. J. Urfi)

While studying the literature on local people's attitudes towards heronries in India, particularly those recorded in the journal of the *Bombay Natural History Society* (BNHS) – a famous natural history institution created more than a century ago by the expatriate British living in India, I discovered a mixed bag of anecdotes and perceptions. Amazingly, from one particular site in South India,

Koonthankulam (BirdLife International 2023c), ornithologists and naturalists had recorded a variety of diverse perceptions from local people over a period of several decades in the first half of the twentieth century. In a reference to Koonthankulam, an observer called Rhenius (1907) in British India first wrote, 'The villagers look on these birds as semi-sacred and will not allow anyone to disturb or molest them, so they return to build there year after year, and have done so for years past.'

Then, almost a half century later, a report by Webb-Peploe (1945), another naturalist in British India, recorded a different view about local peoples attitudes towards it: 'The noise and the smell caused some of the villagers to suggest destroying the nests and driving the birds away some years ago, but the head-men of the village protect them.' A couple of decades later, a note by Wilkinson (1961) from this site seemed to suggest that the locals held the birds in high esteem: 'The headmen of the village still protect the birds and their women-folk spoke with scorn of a village of which they had heard where the people had so ill-treated their birds that "not even a sparrow is to be found there now!"'

As we can see, conflicting views were recorded by various writers pertaining to local people's attitudes towards the birds nesting at Koonthankulam – some seeing it as an opportunity to obtain eggs and meat easily, some finding it a nuisance while others holding the birds in high esteem and praising the locals' positive sentiments towards wildlife.

Kokkare Bellur village: a unique nesting site for Painted Storks and pelicans

In India, it is widely held that the benign, tolerant attitudes of local people towards wildlife, birds and the environment prevalent in some pockets of the country are potentially an asset for conservation of biodiversity. Several examples are known, particularly the practice of some indigenous people in India of protecting small patches of forest known as 'sacred groves'. Members of a rural community in Rajasthan state known as Bishnoi protect wild animals and oppose the hunting of Blackbuck *Antilope cervicapra*, also known as the Indian Antelope. In the context of heronry birds, Kokkare Bellur village, in the South Indian state of Karnataka, where Painted Stork and pelicans nest in large numbers, is rather special (Figure 8.1). For more than a century, Painted Storks, pelicans and other birds have coexisted with local people and generally there has been no molestation of the birds by the local villagers. In recognition of the role of local people at Kokkare Bellur in protecting pelicans, this site has been notified as a Community Reserve by the state forest department.

FIGURE 8.1 A view of Kokkare Bellur village. (Bharat Bhushan Sharma)

Typically, Painted Storks and pelicans start congregating at the site at the end of December or in early January. During the nesting season from January to May when the birds are in residence, they fly in and out of the village nesting ground to forage in the nearby agricultural fields and wetlands. It appears that their principal foraging grounds lie in the vicinity of the nearby Shimsha river. Over the years, felling of nesting trees and intensive fishing and pollution in the surrounding waterbodies have been thought to be the reasons for a decline of heronry birds in this area.

Manu and Jolly (2000) recorded several anecdotes pertaining to local people's positive sentiments towards bird conservation at Kokkare Bellur. The following quotations from their report are pertinent.

> The people of Kokkare Bellur have a long reputation for living with the bird colonies breeding in their village in a harmonious, almost symbiotic way. Though they do not attribute any divine status to the birds, they have always offered them protection, believing that the birds bring them good fortune with regard to the rains and their crops. They are proud of their long association with the birds, nicknamed 'daughters of the

village', and compare them to the local girls who may marry into another village but inevitably return home to deliver and nurse their newborn babies. (Page 9)

The villagers protection of the birds now takes the form of benevolent tolerance for these noisy, smelly annual visitors. Once the season starts there is the ceaseless cacophony of young birds clamouring for food, and the all pervading fishy stench of droppings right there in the villagers' backyards. If the pelicans choose to nest in a tamarind tree, some villagers are even prepared to sacrifice their crops rather than scare off the nesting birds. Village elders actively discourage local children from teasing the birds or stealing their eggs.

... [W]hen a couple of *hakki-pikki* tribals (those who live by hunting birds) came to the village and began climbing trees in search of eggs and chicks, they were shocked to find themselves being arrested by the local Panchayat. Unable to meet the statutory penalty of Rs 100, the tribesmen were tied to a tree in front of the village temple. In another instance in the early '70s, when a Bangladeshi refugee attempted a similar nest-robbing feat, he was caught and locked up for a day in the schoolroom. (Page 10)

Is the protective attitude of the local community really a factor at Kokkare Bellur?

The favourable attitude towards birds at Kokkare Bellur is well documented and much written about, but the question is whether it is for real or just a myth. Nesting birds, storks included, are intensely sensitive to human disturbance (Datta and Pal 1993). Moreover, at many colonial waterbird nesting sites in India the attitudes of local villagers may not always be benign and so this question needs to be asked. But also, in spite of strong anecdotal evidence in favour of locals' protective attitudes, this question should also be asked for other reasons. For instance, it may just be a narrative popularised by amateur birdwatchers and self-styled conservationists who travel from cities to the countryside for the sake of watching birds – in a country torn between diverse social realities and with a huge diversity in terms of class and ethnicity (Urfi 2012).

On a visit made to Kokkare Bellur in January 2019, I recorded 50-plus colonies of Painted Storks in the village. I was struck by the enormous variability of colony sizes and the heights of trees used as a nesting substrate by the Painted Stork during their nesting season, from January to May (Urfi 2021a). Those located on huge trees (about 10 metres in height) with correspondingly large canopies and were quite large, with over 60 nests (Figure 8.2). However, such large colonies were few (only two) in number. Most of the colonies that dotted the village, outside people's homes or in their farmyards, were located on moderate to small-sized trees, in which 3–4 nests were located just a few metres from the ground (Figure 8.3). So, some Painted Stork nests were positioned dangerously low, on trees that could be easily reached by anyone from the ground.

From an ornithological viewpoint, and as we discussed in Chapter 2, losses of eggs and young through predation from the air (chiefly raptors) or the ground (reptiles, mammals) have been an important force in the evolution of avian coloniality and nest-site selection in the context of heronry birds. Exploitation by humans for eggs or meat is also probably a factor, though only realised more recently. As mentioned earlier, all across India generally two types of nesting sites are used by Painted Stork. The first type is trees on islands in marshes, village reservoirs and urban ponds, often on clumps of trees, which can be quite low in height. Painted Stork colonies at Delhi Zoo, Bharatpur and other places are good examples of such a nesting strategy. In fact, at some sites nests can be quite close to the water itself. It is clear that this type of nesting is adaptive, utilising water as a protection against ground predation. The second type of nesting site takes advantage of height to ward

FIGURE 8.2 Painted Storks nesting on a large tree in Kokkare Bellur. (A. J. Urfi)

FIGURE 8.3 Painted Storks nesting on a small tree in Kokkare Bellur. Note the houses in the vicinity of the nesting colony. (A. J. Urfi)

off predation from the ground by situating nests in tall trees. While this may or may not be a perfect solution for protection against all natural predators (say monkeys), at sites like Kokkare Bellur it would certainly deter humans from climbing the trees and extracting the nest contents. In this village, nesting on low trees makes sense only when one considers the locals' benign and non-interfering attitude of leaving the birds alone. It would appear that the birds seem to be cognisant of it and continue to nest there, year after year (Urfi 2021a).

Chapter 9

Conservation

From their thriving nesting colonies in several parts of India, including many in the middle of busy cities, it would seem that Painted Storks adapt to human-altered landscapes quite well. The impression one gets is that Painted Stork are not doing too badly, but as a group of fish-eating birds strongly dependent upon wetlands for their survival, they may be facing a number of challenges. Wetlands all over the world and also in India are under grave threat and so is the biodiversity dependent upon them. Unfortunately, wetlands give the impression of being wastelands, breeding grounds for malaria and mosquitoes and are considered as potential real estate. Developers are often keen to reclaim wetlands and such activities are a major cause of their loss. Thus wetland birds like Painted Stork and others are threatened on account of loss of their natural habitat and other factors such as pesticide and heavy metal pollution (Muralidharan et al. 2016), climate change and invasive fish species.

Invasive fish species

Some of the Indian, exclusively fish-eating birds listed under various IUCN threat categories are listed in Table 9.1. The list includes the Painted Stork (Near Threatened (NT)), Black-necked Stork (NT), White-bellied Heron *Ardea insignis* (Critically Endangered (CR)), Spot-billed Pelican (NT), Dalmatian Pelican *Pelecanus crispus* (Vulnerable (V)), and the Oriental Darter (NT). The general causes of the decline in populations of fish-eating birds include destruction of wetlands, pollution, disturbance and reduction in the quality of foraging habitats by the spread of aquatic weeds like Water Hyacinth *Pontederia crassipes*, among others. To these we can add another dimension, namely the spread of invasive species of fish in Indian waterbodies, an issue which has been much highlighted in recent years (Biju Kumar 2000). Reports suggest that many freshwater bodies in India, local reservoirs, lakes, rivers,

TABLE 9.1 Endangered species of fish-eating birds in India.

Species	Status	Diet	Distribution	Conservation issues
Ciconiidae				
Oriental Stork *Ciconia boyciana*	EN	◑	Winter visitor in northeast India.	Habitat destruction.
Black-necked Stork *Ephippiorhynchus asiaticus*	NT	●	Widespread but nowhere common.	Destruction and degradation of its habitat and overfishing. Pesticide poisoning by consuming infected food.
Lesser Adjutant Stork* *Leptoptilos javanicus*	V	◑	Extensive range across south and South-East Asia.	More partial to wetlands than *L. dubius*. Extensive fishing, loss of foraging habitat and pesticide pollution are major threats.
Greater Adjutant Stork* *Leptoptilos dubius*	EN	◑	In Asia, its three known breeding colonies include those in Assam in India, Tonle Sap lake and in Kulen Promtep Wildllife Sanctuary	Destruction of its foraging grounds as well as nesting habitats.
Painted Stork* *Mycteria leucocephala*	NT	●	Widespread throughout the Indian subcontinent.	Loss of foraging habitat, pollution.
Anhingidae				
Oriental Darter* *Anhinga melanogaster*	NT	●	Widespread	Overfishing, loss of wetlands. Hunting in some localities.
Pelecanidae				
Spot-billed Pelican* *Pelecanus philippensis*	NT	●	Found in well-watered tracts of almost all of India.	Loss of foraging habitats, overfishing, stealing of its eggs.
Dalmatian Pelican* *Pelecanus crispus*	V	●	Wide breeding range across Europe, central Asia. Prefers inland freshwater waterbodies but also occasionally in lagoons.	Extensive fishing in its foraging grounds, loss of wetlands and also poaching.
Ardeidae				
White-bellied Heron *Ardea insignis*	CR	●	Has a narrow distribution in eastern India: mainly reported from Assam, West Bengal, Sikkim, Arunachal Pradesh; also from Nepal, Bhutan, Bangladesh, central Myanmar and a few other places.	Widespread loss, degradation and disturbance of forest and wetlands. Due to extensive fishing its prey base is declining.

Species	Status	Diet	Distribution	Conservation issues
Threskiornithidae				
Black-headed Ibis* *Threskiornis melanocephalus*	NT	◑	Widespread	Wetland loss and pollution, poaching, hunting.
Accipitridae				
Pallas's Fish Eagle *Haliaeetus leucoryphus*	V	●	Main breeding populations are believed to be in China, Mongolia and the Indian subcontinent. In India reported from larger rivers and wetlands in the north and northeast.	Being at the apex of food pyramids in aquatic ecosystems, first to be affected by impacts lower in food chain. Habitat loss, spread of Water Hyacinth, felling of large trees near wetlands, agricultural pesticides.
Lesser Fish Eagle *Ichthyophaga humilis*	NT	●	Himalayan foothills, northeast India with small populations in Karnataka and Kerala. Said to be destructive to trout fisheries.	Loss of forest habitats along rivers, increasing human disturbance, pesticide contamination.
Grey-headed Fish Eagle *Haliaeetus ichthyaetus*	NT	●	Prefers comparatively sluggish rivers and streams flowing through undisturbed forests.	Loss of undisturbed wetlands, overfishing, siltation, pollution and persecution. Pesticide contamination and eggshell thinning.
Heliornithidae				
Masked Finfoot *Heliopais personatus*	EN	◑	Northeastern India, Bangladesh, through Myanmar, Thailand, Cambodia, Laos, Vietnam, Malaysia, Sumatra, Java and Indonesia. A denizen of dense lowland forest pools and streams, including coastal wetlands and mangroves.	Habitat destruction, poaching
Laridae				
Black-bellied Tern* *Sterna acuticauda*	EN	●	Widespread on long stretches of placid waters and resting on river islands and sandbanks.	Loss of nesting and foraging habitat.
Indian Skimmer* *Rynchops albicollis*	V	●	Found on larger rivers across the Indian subcontinent and some countries in South-East Asia.	Destruction of nesting habitats and foraging grounds due to anthropogenic pressures on land, pollution, spread of aquatic weeds.

(continued)

TABLE 9.1 Endangered species of fish-eating birds from India. (*continued*)

Species	Status	Diet	Distribution	Conservation issues
Alcedinidae				
Blyth's Kingfisher *Alcedo hercules*	NT	●	Rare bird restricted to northeast India.	Habitat destruction and fragmentation.
Brown-winged Kingfisher *Pelargopsis amauroptera*	NT	●	Mangroves, creeks, tidal rivers on the eastern coast of India; some countries in South-East Asia.	Development in coastal regions.

Source: Rahmani (2012)

Abbreviations used: EN, Endangered; NT, Near Threatened; V, Vulnerable; CR, Critically Endangered. Symbols used: ● exclusively fish diet, ◐ largely carnivorous diet (reptiles, birds, carrion), includes fish, * colonially nesting bird

streams and wetlands, are infested by aggressive, alien species of fish, which may have accidentally got into Indian waters, and may interfere with the food webs in wetlands. One such species is the African Catfish *Clarias gariepinus*, which grows quite large in size and proliferates rapidly, dislodging many native species of fish in the process, especially those that may constitute the natural diet of fish-eating birds.

The presence of exotic fish species in Indian waters is by no means a new phenomenon. For instance, tilapia were introduced in India during the 1950s (Jhingran 1982) for the purpose of aquaculture, but have now spread all around the country. In Chapter 7, tilapia were mentioned when discussing the information available on the diet of the Painted Stork. Tilapia were missing from a detailed study of gut-content analysis conducted during the 1960s on the same population of Painted Stork (Desai et al. 1974). This gives us an idea how rapidly this species has spread to become an integral component of aquatic food webs. Whether this in itself is good or bad for a fish-eating bird is difficult to say at this point; yet these studies imply that tilapia are present in sufficiently high numbers that they may well have had some impact on the composition of the community of freshwater fish. There is already a widespread fear of invasive fish species producing alterations in the structure of native fish communities in Indian waters (Khan and Panikkar 2009).

From information available in the literature it is clear that the African Catfish is a highly aggressive species, a fast breeder (like tilapia) and preys upon existing fish, quickly rendering the waterbody devoid of other species (Biju Kumar 2000). Due to its size and aggressive behaviour it is unlikely to be a part of the diet of fish-eating birds such as the Painted Stork. Given the renewed interest in aquatic birds and their ecosystem impacts (Kerekes and Pollard 1994; Hanson and Kerekes 2006) and also the greater appreciation today of the ecosystem services provided by waterbirds (Green and Elmberg 2013), further investigations in this area are warranted.

During field surveys that I undertook in North India in 2015, I came across several waterbodies in the state of Rajasthan that seemed to be heavily infested by African Catfish. Kishangarh Fort pond, for instance, located approximately 30 kilometres from the city of Ajmer, had large numbers of extremely big catfish, later identified as African Catfish, which congregated at the shore whenever locals threw food at them (Figure 9.1). The large tank looked completely devoid of waterbirds, which seemed strange because my visit was in December – a time of the year when migratory waterbirds would generally be found in large numbers at wetlands in North India. However, other reasons could also be responsible for the absence of waterbirds here, besides the presence of African Catfish.

Is the Painted Stork declining?

The primary thing in conservation is numbers. Sizes of populations matter. If numbers of a particular species decline then it means that it is in trouble. But simply saying 'numbers' may be a bit vague and so we need concrete figures for a region or country. In other words, numbers or population sizes need to be viewed in a proper context, and before we proceed further it may be worthwhile to examine how and in what manner numbers, so useful for research and development of conservation policy, are arrived at. We also need to look at what efforts go into establishing numbers of wildlife populations, especially at the scales required.

In the 1990s the Wildfowl and Wetlands Trust (WWT) and Asian Wetland Bureau (AWB) launched an important programme known as the Asian Waterbird Census (AWC) in an attempt to get some idea about numbers of different species of waterbirds wintering in Asia (Urfi et al. 2005). It should be borne in mind that not all species are equally distributed or abundant. For instance, some waterfowl such as migratory Pintail *Anas acuta*, Mallard *Anas platyrhynchos*, Shoveler *Anas clypeata* and others are much more numerous while others such as, say, Gadwall *Anas strepera* tend to be much fewer on their wintering grounds in India – something that could be accounted for by several factors. The data supplied for the AWC were gathered by amateur birdwatchers, volunteering their time and effort to provide information, and there were some imperfections in the data. Furthermore, waders like Painted

FIGURE 9.1 African Catfish *Clarias gariepinus* in a water storage tank in Rajasthan. The species of fish in this picture was identified by Professor Abebe Getahun, Professor of Ichthyology, Department of Zoology, University of Addis Ababa, Ethiopia. (A. J. Urfi)

Storks are likely to be present at smaller ponds and marshes in the countryside and not just at larger ponds and reservoirs where many volunteer birdwatchers congregate, for logistical reasons, to count birds (Sundar 2006). While there may have been considerable under-recording of waterbird populations, the

AWC allowed estimates for various species of waterbirds in India to be made, which set the foundation for more census exercises later. According to BirdLife International (2023a) the population size of the Painted Stork across its range is estimated to be 16,000–24,000, with a decreasing trend. Its extent of occurrence (breeding/resident) is estimated to be around 6,090,000 km^2.

Site-specific estimates: Keoladeo Ghana National Park

Population size fluctuations from select sites can provide a more meaningful picture, which is often easier to understand. From India we have long-term population data trends for Painted Stork from only a few sites, of which Keoladeo Ghana National Park (KDGNP) and Delhi Zoo are prime examples. Let us take KDGNP first. This beautiful marsh looked enchanting once upon a time. In its heyday, this picturesque Ramsar site featured in the Hollywood film *Siddhartha* (based on a Herman Hesse novel of the same name), but today it is a parched version of its former self. Among the park's distinctive features was the presence of extensive heronries. Since this site has been closely monitored by bird lovers, conservationists and environmental activists over the years, we have some idea about the fluctuations its Painted Stork populations. At the turn of the previous century, the number of breeding pairs of Painted Storks in the extensive park was estimated conservatively to be around 4,000 (Ali 1953; Sankhala 1990). If all the other species of colonial waterbirds nesting in the park were included then the total number of breeding pairs would be in excess of 10,000. However, since the early 2010s numbers of breeding pairs of all manner of colonial waterbirds recorded have declined considerably. Recent observations at KDGNP gave an impression that the total number of Painted Stork nests is not more than a few hundred.

Another dimension to the decline of Painted Stork at KDGNP is that while in the early decades of the twentieth century nesting colonies were found in several blocks, in recent years their spatial distribution in the park has also shrunk, and they are now restricted to just a few blocks (Figure 9.2). So, the most obvious question is what has caused the decline of Painted Storks nesting in Bharatpur? Unfortunately, this is not an easy question to answer as there may be several explanations related to factors operating inside and outside the park. There is a view that the flow of water inside the park after the monsoon from the rivers (Gambhir) nearby has been tightly regulated by the authorities. Some social scientists who have examined this problem feel that water was diverted from the park to agricultural activities in other areas, causing the wetland ecosystem in the park to collapse (Chouhan 2006).

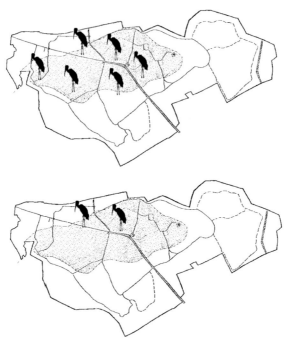

FIGURE 9.2 Sketch map to show how Keoladeo Ghana park is bisected into a number of blocks created by elevated pathways. (Top) Prior to the 1980s, Painted Storks and other heronry birds occupied almost all of the blocks of Keoladeo Ghana National Park for nesting purposes. (Below) In recent years, Painted Storks have nested in only a few blocks (B and D) of the park.

Painted Stork populations at other sites

The changes happening at different Painted Stork nesting sites in India have been reviewed in Urfi (2011a) and only the main points are mentioned here. Regarding Kokkare Bellur, reports in 1977 by Neginhal indicate that the village's Painted Stork colony was in a phase of expansion at the time of his visit. According to local estimates, more than 1,000 pairs of pelicans were nesting at Kokkare Bellur in the 1980s. But a decade later the number of breeding pairs had come down to 160. According to another contemporaneous report, Painted Storks at the village ranged between 850 and 900 and pelicans between 300 and 350 nesting pairs. The decline was attributed to tree felling, hunting, decline in food supplies and pesticide poisoning (Manu and Jolly 2000).

However, at the Piele Gardens and nearby Ghogha area in the city of Bhavnagar the trend seems to be the reverse. According to Parasharya and Naik (1990) the total number of Painted Stork breeding pairs recorded from Piele

Gardens were 96 and 70 in 1980 and 1981. Later, while the Ghogha colony disappeared, those in Piele Gardens grew in size. According to local sources, there has been an increase in the numbers of Painted Stork at Bhavnagar's city gardens (although this could not be verified at the time of writing this book). While it is held that the local municipal authorities have worked hard to maintain the city gardens and also ensure that visitors do not disturb the birds, what are we to make of these trends, if verified to be correct? Are more Painted Storks nesting in Bhavnagar because overall Painted Storks are doing well and are becoming abundant? Or is it because in the country, as their natural breeding habitats are being lost, they are being forced to flock to the safe confines of gardens in cities, such as the Piele gardens of Bhavnagar, which afford some degree of protection and also suitable nesting habitat?

The case of the Delhi Zoo colony is somewhat similar. In 1960 close to 500 Painted Storks were recorded in the zoo premises, and over the years, though there have been annual fluctuations in numbers of nests (explained by the performance of the annual monsoon rains), overall there seems to be a degree of stability in this population. This is despite the fact that the environment around Delhi Zoo has deteriorated considerably (see Chapter 7). There has, for example, been an increasing trend of pollution of the River Yamuna, human encroachment on the floodplains, thereby reducing the amount of foraging habitat and increases in human activity, noise and disturbance. This can only mean that the stability of the population is due to their safety in the zoo premises (Urfi 2020).

Conservation of heronry birds: role of environmental education

The importance of sensitising the general public to issues pertaining to ecology and conservation of wetlands and their biodiversity cannot be overestimated. It is imperative that efforts are made to impart information to the public via a number of methods and by employing appropriate strategies in different situations, sometimes using novel methods of outreach towards communities (Box 9.1). The idea is to engage the attention of the visitor who comes to a bird sanctuary or stands in front of a heronry and spends some time there. The objective in environmental education (EE) is to ensure that a visit does not remain just an outing or a picnic but also has a take-home message which can be transmitted to others. In EE terms this is known as the 'multiplier effect' – one person tells another and so on until a wider section of the general public is informed with facts. EE is also the process of triggering a degree of curiosity about the natural world and its habitats. The end result is, one hopes, that

BOX 9.1 Hargila Army – an all-female conservation group in northeast India named after the Greater Adjutant Stork

In the Brahmaputra valley of Assam in northeastern India, the Greater Adjutant Stork is locally known as *Hargila*, which means 'bone swallower'– an epithet earned from its habit of scavenging and feeding on carcasses. In Assam where the storks nest colonially in privately owned trees within densely populated villages, tree owners cut them down to prevent rotten food and excreta of this carnivorous bird from falling into their gardens – a notable cause of the stork's decline.

In 2007 a conservation project involving community development, capacity building within local communities, and education and outreach work to interlink storks with local traditions and cultures was initiated by local environmental activists. A rural women's conservation group named the Hargila Army was instituted and strong feelings of pride and ownership for the storks by the villagers generated (Barman et al. 2020). It is noteworthy that in this programme cash incentives for nest protection were deliberately avoided. Instead, schemes that indirectly contributed to the livelihoods of nest-tree owners and other villagers were introduced. The activities of local women included rehabilitation of

Local fisherwomen, volunteers for the Hargila Army, participating in an awareness event to mark World Wetlands Day in Pandona Beel (wetland) in Assam. Note the unique hats shaped like Greater Adjutant Stork heads. (Dr Purnima Devi Barman)

injured or fallen nestlings, creation of artificial nesting platforms and the launch of campaigns to prevent trees which are suitable nesting substrate for Adjutant Storks from being felled. Other activities included incorporating the Adjutant into local folk songs, traditions and cultural festivals. For example, villagers give Adjutants that are about to lay eggs a baby shower, using the same rituals as for expecting Assamese women. Women of the Hargila Army also weave images of the Greater Adjutant into their fabrics, spreading awareness about conservation while generating income for their families. The success of the conservation programme is shown in the increase in the number of nesting colonies in some of the villages in the area. A change in mindset of the nest-tree owners towards keeping their trees and in developing a positive attitude concerning Greater Adjutant Storks has been the key to this conservation project.

a sizeable proportion of the general public is aware of the need to conserve biodiversity and can rally behind causes that seek to put pressure on governments to conserve habitats and even participate in biodiversity monitoring events. However, this is easier said than done, partly because most responsible agencies lack sufficient funds and resources to sustain EE programmes. An important question is how to devise EE strategies for diverse kinds of visitors. In India this is a serious challenge because visitors to parks, sanctuaries and zoos are an extremely heterogenous group who speak different languages and come from various social and educational backgrounds (the most obvious being the rural–urban divide).

A heronry that at a first glance looks like a wall of nests presents a unique opportunity for EE (Figure 9.3; Urfi and Nareshwar (1988)). It is a place where one gets to see birds while they are reproducing – the most important phase in their life. Standing close to a heronry with nests of Painted Storks, ibises, herons, egrets and cormorants, one sees them engaged in courtship rituals, mating and nest-building, then guarding and feeding their chicks, and watching them grow. One also sees the dependence of birds on a variety of resources, not only at the heronry site but outside it in the wetlands in the vicinity, where the nesting birds make regular trips to gather food for the chicks. The easiest and simplest method for EE is to install information panels at the site, with at least one graphic and a short message, in a language that can be read by most people in the area. In the context of India this means that, besides English, the text has to be in a local language (Hindi in most parts of North India and other languages in other regions). Installation of information

FIGURE 9.3 A heronry (which at a first glance looks like a wall of nests) presents a unique opportunity for environmental education. This picture shows a group of students on an educational tour standing near the entrance of Koonthankulam Bird Sanctuary in Tamil Nadu state, South India, where the extensive heronries of Painted Stork sand other waterbirds are a highlight. (M. Mahendiran)

panels/waysides is a complex process which involves research into selecting the appropriate material to withstand the harsh tropical sun and vagaries of the weather, scientifically correct and up to date information, and relevant graphics. Information can be also circulated by way of leaflets and brochures, and online using digital materials. A slightly more advanced method that caters for someone who has taken the trouble to visit the site, is to provide an 'orientation centre' in which information is provided in an entertaining manner through audiovisual methods and dioramas. At many bird sanctuaries and important bird areas in India some efforts have been made in this direction. For instance, at KDGNP a nature interpretation centre named after Dr Salim Ali has been set up which has dioramas of Painted Storks nesting (Figure 9.4). Orientation centres have also been made at Sultanpur National Park near Delhi and Kokkare Bellur village.

Among the target groups for EE purposes, besides students and visitors to sanctuaries, are local communities and the need for involving them in protection is increasingly being realised. A case in point is KDGNP, where a series of significant events in the 1980s and 1990s, illustrating the unique

FIGURE 9.4 A view of the Salim Ali Interpretation Centre at Keoladeo Ghana National Park with dioramas of heronry birds. (A. J. Urfi)

conservation problems faced in developing countries (Urfi 2021b), created space for a fresh dialogue on environmental conservation in India. Simply put, the story is as follows. The world-famous park, besides being a place of high biodiversity and conservation significance, also happens to be the only grazing ground for thousands of cattle belonging to farmers from the nearby villages. In the arid, hostile wilderness of the desert state of Rajasthan, competition for scarce resources is much heightened. In the 1980s some well-meaning scientists, ornithologists and conservationists – all erudite and highly educated people, headed by a team from Bombay Natural History Society under the stewardship of Dr Salim Ali – appealed to the central government to enforce a ban on cattle grazing in an effort to make Keoladeo an inviolate space for birds, particularly for the highly endangered Siberian Crane which used to winter in the park. For centuries the villagers had been grazing their livestock in the park and there had been no objection from the local ruler of the princely state of Bharatpur, whose ancestors had created the park as a private hunting preserve. But now with this ban, enforced with all might of the state, the villagers were incensed. There were numerous clashes between local people resisting the ban and the police and park officials in which several people died. Questions began to be asked, including 'Is Bharatpur for birds only?'– similar to the question that was asked in the context of Project Tiger

FIGURE 9.5 In the 1980s a cattle grazing controversy erupted at Keoladeo Ghana National Park. One newspaper headline read: 'Is Bharatpur for birds only?', raising the age-old issue of biological conservation versus human land-use. (A. J. Urfi)

a couple of decades earlier (Fig 9.5). To cut a long story short, after the ban on grazing was imposed, a common invasive grass, *Paspalum vaginatum*, which had been previously kept in check by cattle, now grew and choked the wetland – rendering it useless as a habitat for migratory waterfowl, including the precious Siberian Crane and the heronry birds (Ali and Vijayan 1983). In the end, the authorities had no choice but to lift the ban on grazing.

Monitoring populations for urbanisation and climate change

In order to understand how urbanisation, climate change and other issues influence bird populations, long-term data is a necessary pre-requisite (Crick and Sparks 1999; Crick 2004). It is only with the help of long-term data sets that it becomes possible to grasp the role of density-dependent and density-independent factors in population regulation. One textbook example is how long-term data on Grey Heron populations in the United Kingdom helped us to understand the role that weather patterns (harsh winters in this case) play in impacting bird populations (British Trust for Ornithology 2022). The recent breakthroughs in understanding how global climate change can influence bird migration (say arrival dates) has been possible due to analysis of long-term data sets in Europe. So having a system in place to gather long-term data on birds is important for understanding the effects of environmental change on biodiversity. There is a strong need for conservation monitoring programmes that can be sustained over long periods, with mechanisms in place to ensure regular data gathering by trained personnel (Urfi 2020). Having said this, and despite the known benefits of long-term population monitoring, most agencies are reluctant to invest given that it requires a continuous supply of funds, commitment to science, strict adherence to a protocol and plenty of coordination activities on the part of the organisations involved. Box 9.4 offers an overview of two international bird population monitoring programmes.

Rescue and rehabilitation initiatives

While EE has a significant role to play in conservation, there are many other areas of engagement and activity. In India several individual naturalists and bird lovers have taken it upon themselves to initiate community-level efforts to conserve stork colonies and rehabilitate chicks that fall from the trees, sometimes with the help of the state's forest department. At Koonthankulam one such initiative is underway and special pens have been created to house the rescued chicks (Figure 9.6). At Kokkare Bellur some good samaritans have

FIGURE 9.6 Efforts to rescue and rehabilitate Painted Stork chicks displaced from nests underway at Koonthagulam Bird Sanctuary. (M. Mahendiran)

been at work rescuing fallen nestlings, which otherwise often become prey to cats, dogs and snakes (Subramanya and Manu 1996). The initiative was begun by K. Manu, a birdwatcher from Mysuru, who also created an environmental organisation (Mysore Amateur Naturalists) as a platform for his conservation activities. Manu in turn inspired a local resident of the village, a farmer by the name of B. Linge Gowda, who for several years has been looking after chicks that fall from their nests and tending to them until they are big enough take wing and fly off. For this purpose, he built a pen to house the young birds and fed them fish; all these activities were undertaken at his own expense, for the love of the birds. Linge Gowda inspired several local village youths to join his cause and saves dozens of stork and pelican nestlings each year. Thanks to the efforts of these two – Manu and Gowda – 'Hejjarle Balaga', a community initiative, was started by locals for the conservation of birds in the village. In coordination with the state government's forest department a number of other initiatives have been started, such as nature education programmes, taking up the issue of the construction of a power line through the village which would have resulted in the death of many birds and more.

A nest-monitoring programme for Painted Stork colonies

Over the years, the need for starting a nest-monitoring programme for Indian heronries has been advocated (Urfi 2020), with monitoring of the Delhi Zoo Painted Stork population as a pilot programme. This colony, as we have discussed before, offers some unique advantages, particularly the ease with which observations can be made. The standard method involves recording information on individual nests, preferably marked or readily identified.

BOX 9.2 Importance of museum collections in conservation

Several nesting parameters are crucial to understanding patterns of decline of bird populations. While nesting success is an important parameter, clutch size information can be useful too. Rodgers (1990) obtained zoological data from museum collections to analyse historical temporal patterns of breeding and clutch size of Wood Stork populations in Florida. Did a historical decrease in clutch size contribute to their population decrease? Information on 203 museum tags that accompanied Wood Stork clutches collected during 1875–1938 and in 1967 was analysed to establish breeding chronology and clutch size in Florida. Although the 1885–1894 period exhibited significantly larger clutches, no overall trend was detected in clutch size during 1875–1967. Modal and mean clutch sizes during the late 1800s and early 1900s were similar to those of the 1980s. However, that such a study could be undertaken underscores the importance of museum collections of eggs and skins to conservation.

(Box 9.3). The software for estimating nesting success is based on the mark and recapture method, a well-known method in population ecology. It is crucial that there are no mistakes in following the fate of individual nests on successive field visits. In this regard, while marking the nests is highly advantageous it may not be always possible. For instance, the nests might be located on islands which are difficult to reach. Often permits by the forest department are not granted making it impossible to mark the nests. However, this problem can be overcome by taking photographs of the colonies and marking the nests on printouts or images, as described in Tiwary and Urfi (2016). Undertaking monitoring exercises along scientific lines is discussed in Box 9.4.

BOX 9.3 Monitoring stork nests in the Florida Everglades: a first-hand account

For years I had read about the 'rivers of grass' and its denizens, the Wood Stork, other colonial waterbirds and alligators, so when it came to my choice about selecting an institution to base myself as a visiting scholar from India, I was pretty sure it would be near the Everglades in Florida. In Miami, gateway to the Everglades, I was greeted by the coordinator of the field research programme for the Department of Wildlife, University of Florida, Gainesville and he initiated me into the rigours of field work. A typical field day started early. Getting up at 5 a.m., picking up the air-boats, fixing them on trailers and reaching the shore. Then, getting the boats with their huge propellers (powered by an aircraft engine) into the water and heading off to the islands where the nesting colonies were. On my first field day, when the boat was parked and entangled in the dense vegetation close to the colony, I began wondering what the next step would be. Then it dawned upon me that the only way of getting to the nests up-close would be by wading into waist-deep mud and walking to them. But just to be sure I decided to check this out with John, the coordinator of the field research programme, and his solemn reply was, 'Sadly, yes!' And so there we were, mucking around in the Everglades.

The task at hand was checking nests of Wood Stork, Anhinga *Anhinga anhinga*, Great White Egret *Ardea alba*, White Ibis *Eudocimus albus* and other species. To do this the team would move along a flagged path, stopping on the way at nests that had numbers written on strips of tape stuck on nearby branches. We would check the nest contents using a mirror mounted on a long metal pole and shout out the contents, which another team member would dutifully record in her diary. 'Nest number 10: two eggs. Number 23: three chicks. Number 4: empty...' and so on.

Walking the flagged path mostly in waist-deep water (and sometimes with alligators swimming in the vicinity) normally took 3–4 hours and by noon we were ready to head back home, exhausted.

The bird colonies were widely scattered across the Everglades and due to having to trudge through them, one gets to know only a small sample; a complete census is difficult. However, Prof. Peter Frederick, director of

The author examining a heron chick in the Florida Everglades. (John Simon)

the research programme, was also looking at ways to census the colonies completely by using aerial photography. One day, one of the team members offered to show me how the colonies looked from above, while practising her census method (and also her flying, as I discovered subsequently). I leapt at the invitation, but must confess I felt a surge of adrenalin as the door-less single-engine aircraft took off and the distance between the ground and us mortal human beings became larger and larger. But having adjusted to it, I relaxed and began to enjoy the bird's-eye-view of the colonies, which first looked like tiny white specks but became recognisable as we descended nearer to them.

One of our tasks while out in the field was to sample some dorsal feathers from the heron chicks for the analysis of heavy metals and pesticides. One of the biggest threats to the Everglades and its birds comes from aquatic pollution. Heavy metals such as lead, mercury, cadmium, and pesticides act in insidious ways. They reach the waters through surface runoff and leaching, and, once in the water, they travel along the food chain. At each step, their proportions increase exponentially through biomagnification in the bodies of organisms who have eaten that food. Inside the bodies of top-level predators like fish-eating birds these toxins play havoc and start interfering with the normal metabolism of the birds and eventually affect their reproduction and nesting, leading to population declines.

Extracted from Urfi (2006b)

BOX 9.4 Monitoring nests: international programmes involving volunteers

Showcasing the nesting phase – a significant chapter in the lifecycle of birds – heronries offer a great opportunity for research and conservation monitoring. Unlike bird-counting programmes, nest-monitoring places emphasis on nests. Fortunately, nests are stationary structures that cannot fly off, unlike their makers! The focus in nest-monitoring programmes is on the contents of the nest, eggs and nestlings, alongside information on parent birds in attendance. By repeated visits to the same nest, one obtains estimates of nest success – an important parameter in conservation biology that helps accurately understand how birds respond to change in their habitat. Some agencies have started to monitor nest success using specialised software and volunteer effort. The NestWatch programme of the Cornell Lab of Ornithology (Cornell Lab of Ornithology

2023), which also runs another popular programme called eBird, is one such case. Another important example is the Heronries Census (operational since 1928) by the British Trust for Ornithology (2022).

Eggs in a heron's nest. Data on eggs and nestlings, along with information on parent birds, is important for conservation. (A. J. Urfi)

Outlook on the future

Some years ago while on a study tour in Bharatpur, I noticed a Painted Stork behaving very strangely. Feeding in a marsh, it looked quite normal at first but when it took to the wing it made a wavering flight. Vainly trying to get a foothold on a branch it lost its balance and crashed to the ground. The bird was visibly stressed and seemed to lack motor coordination, and I wondered what could account for its strange behaviour. Could it simply be old age? Or was it poisoning of some sort. In the agricultural fields around Bharatpur bird sanctuary pesticides (like Endosulfan, Malathion, DDT and so on) are extensively used; possibly the bird had ingested contaminated food which had affected its neuro-muscular coordination. Pesticides act in insidious ways, such as blocking the enzymes at the synaptic junctions and preventing the smooth transfer of nerve impulses, thereby affecting movement. While I had no opportunity to ascertain the exact cause of the bird's strange behaviour my thoughts were diverted to how toxins travel along aquatic food chains and affect the top-level carnivorous (fish-eating) predators like Painted Stork.

The study of food chains in wetland ecosystems in India is very important, as wetlands are being lost to land encroachment, pollution and other factors across much of Asia. Yet several crucial aspects of their ecology – especially the role of birds in ecosystem functioning, the impact of pesticides and the presence of invasive species of fish like African Catfish – remain largely unexplored. Equally significant in the context of foraging is the relationship of aquatic food cycles with the monsoon. This is an important topic given the concerns over global climate change. Hence, the creation of databases, as part of initiating a long-term population monitoring programme and nesting success programme, are likely to be useful for conservation.

In the preceding chapters I have attempted to showcase the Painted Stork as a model to explore some interesting problems in ecology and conservation. The question of what good can come from a study of this species was raised at the outset, and I have discussed how this large and eye-catching colonially nesting wetland bird – which builds its 'rude' platform nests on trees in villages, inside protected areas and on vegetation growing on islands in the middle of water tanks in both rural and urban locations – offers us a unique opportunity to consider issues related to ecology and the impacts of climate change and urbanization on biodiversity loss.

Seen from above, the nesting colonies of Painted Stork and other heronry birds, over a subcontinental scale, may look very remarkable indeed. They could appear as clumps of dots across a large triangular landmass, several thousand kilometres in length, with the sea on two sides, and encompassing widely different climatic regimes. Innovative approaches to analysing the distribution patterns of objects have provided interesting insights about distribution patterns of bird colonies at different geographical scales. For instance, Jovani and Tella (2007) analysed the distribution of nests of White Stork in and around Doñana National Park in south-west Spain. They found a fractal pattern in each of the four study years. Thus, although storks were locally highly clumped, with even tens of nests in a single tree on some occasions, the population was not actually structured in colonies (a simple clustered distribution) as previously thought. Rather, it was organised in a continuous hierarchical set of clusters within clusters across scales – and with these clusters lacking the commonly assumed characteristic mean size. These quantitative solutions to previously perceived scaling problems can aid in improving our understanding of the evolution of bird coloniality.

Furthermore, the sheer enormity of the geographical scale over which heronries are located across the Indian subcontinent is itself breathtaking. Painted Stork populations in North India are separated from those in the south by more than 2° latitude. This means that they are at two extreme ends of the spectrum, experiencing hugely divergent climates and monsoon patterns. In a recent study, the relationship between local bioclimatic conditions and the morphological variations seen in wild Painted Storks from two separate ends of the country, separated by a distance of over 2,000km, were examined (Mahendiran et al. 2022). There was evidence of operation of eco-geographic rules (Bergmann's and Allen's rules) between the regions, resulting in significant body-size variation and sexual dimorphism in Painted Stork from the two different clusters. The distribution of Painted Stork across such a wide geographical area also raises the question of genetic diversity. Since the study of genetic markers is likely to be extremely useful, we have carried out a preliminary study of DNA microsatellite markers in this species (Sharma et al. 2014, 2017a,b) which may be a foundation for further work on population genetics of Painted Stork across India.

Colonial nesting is a defining feature of Painted Stork and other heronry birds. But there still seems to be controversy about the causes of coloniality (Brown and Brown 2001). While the Painted Stork is almost always colonial, one does occasionally come across single nests on a tree in the middle of a marsh or a flooded agricultural field. The question of whether birds can be colonial in some situations while nesting singly in others (i.e. facultative and obligatory) has arisen. More studies on coloniality in Painted Stork and other species on will be useful.

Further studies on nesting ecology, and investigations on the role of various biotic and abiotic factors influencing nesting success in Painted Stork are also required. Our research on colonies in North India has shown the significance of abiotic factors like environmental temperatures during the period when there are chicks in the nest (Tiwary and Urfi 2016a). Since minimum temperature emerged as a significant variable influencing nesting success, and given that in North India temperatures can get as low as 4°C in the months of December/January and possibly be an important cause of mortality in newly born chicks which lack the protective feathering, it will be interesting to conduct more studies, particularly in South India where temperatures are less extreme and the climate more benign.

The Delhi Zoo colony that has been discussed at several points in this book is a good example of a heronry in an urban setting. Such colonies serve as indicators of the changes going on in the wider environment – namely, the spread of urbanisation and the increasing extent of built-up areas. For instance, at the colonies, nests built, yearly nesting success estimates and other parameters may fluctuate as a consequence of changes in the nesting and foraging habitat but we can know these things for sure only if a carefully designed nest-monitoring programme is in place. What parameters of the population or individuals to monitor, and how to do so, is the turf of biologists, and I strongly advocate a heronry monitoring programme with Delhi Zoo as a pilot project (Urfi 2020).

Lastly, one of the themes discussed in this book is people's attitudes towards biodiversity and the question of how heronry birds adapt to human presence. In this regard, the Painted Stork and pelican nesting colonies on trees in Kokkare Bellur village present an interesting case (Urfi 2021a). Here, it seems the birds have adapted to living in close proximity to humans aided by the fact that the local villagers appear to have a benign attitude towards them. Yet in the literature we see a variety of views on heronry birds. Given that most narratives about people's attitudes towards wildlife are authored by urban, well-educated environmental activists (Urfi 2012), whose relationship with rural spaces might be biased, more studies conducted by anthropologists (and not necessarily biologists interested in conservation) may be useful. On the conservation front, the efforts underway by local activists and birdwatchers such as launching local rehabilitation programmes, caring for displaced nestlings, education and awareness initiatives, legislation and enforcing a strict conservation code are standard. To this we must add heronry monitoring, as it is only through maintaining a database with scientifically collected data that we can make informed decisions for conservation.

Appendices

I Initiation of nesting period of Painted Stork at sites across India[a]. See Box 6.3 for a map of these sites

Site name	State	Initiation of nesting period	Reference
North India and western India			
Delhi Zoo	Delhi	End August	Urfi 2011a
Bijana	Haryana	Mid-August	Urfi 2011a
Sultanpur NP	Haryana	Mid-August	Urfi et al. 2007
Govardhan	Uttar Pradesh	September	Hume and Oates 1890 (in Urfi 2011a)
Chaata	Uttar Pradesh	August to September	Tiwary et al. 2014; Tiwary and Urfi 2016a
Khanpur	Uttar Pradesh	August to September	Tiwary et al. 2014; Tiwary and Urfi 2016a
Hazratgunj (town centre), Lucknow	Uttar Pradesh	August to September	Tiwary et al. 2014
Kandhi, Kanpur Dehat	Uttar Pradesh	August to September	Tiwary et al. 2014
Keoladeo Ghana NP	Rajasthan	End August	Ali and Vijayan 1983
Man Marodi Island, Gulf of Kutch	Gujarat	End August	Urfi 2003a
Bhavnagar City Gardens	Gujarat	End August	Parasharya and Naik 1990
Munjasar Tank	Gujarat	Early September	Urfi 2011a
Chikhaliya	Gujarat	September	Urfi 2011a
Seelaj	Gujarat	End August	Urfi 2011a
Central and South India			
Bhadalwadi	Maharashtra	Last week March	Pande 2006
Karanji Tank	Karnataka	January/February to May/June	Jamgaonkar et al. 1994; A. J. Urfi (pers. obs.)
Kokkre Halle	Karnataka	January/February to May/June	Jamgaonkar et al. 1994; A. J. Urfi (pers. obs.)
Ranganthitoo	Karnataka	Jan/February to May/June	Jamgaonkar et al. 1994

Site name	State	Initiation of nesting period	Reference
Kokkare Bellur	Karnataka	December/January to June	Neginhal 1977; Manu and Jolly 2000; A. J. Urfi (pers. obs.)
Kaggaladu	Karnataka	November to May	Urfi 2011a
Veerapura	Andhra Pradesh	Last week January	Bhat et al. 1991
Chintapalli	Andhra Pradesh	January to June/July	Urfi 2011a
Chinna Maduru	Andhra Pradesh	January to June/July	Urfi 2011a
Telineelapuram	Andhra Pradesh	Mid-May	Nagulu and Rao 1983
Vedurupattu	Andhra Pradesh	November/December–May	Urfi 2011a, Pattanaik et al. 2008, Sharma and Raghavaiah 2000.
Koonthankulam	Tamil Nadu	End March/April (also end December)	Rhenius 1907; Webb-Peploe 1945; Wilkinson 1961
Kalpakkam	Tamil Nadu	January/February	Rajaram (1999) in Urfi 2011a
Vedanthangal	Tamil Nadu	Early January	Paulraj and Kondas 1987; Paulraj 1988

[a] In those cases where nesting date was not mentioned it was estimated from age or size of the nestlings.

II Population genetics

The wide geographical spread of the Painted Stork, its capacity to disperse far away from natal sites and the seeming concentration of its nesting colonies in two broad clusters across the Indian subcontinent warrants further studies along biogeographical lines. In this regard molecular genetic studies are likely to be extremely useful besides providing important baseline data for conservation of this species. Among all the four species of *Mycteria*, DNA microsatellite markers – a powerful tool for molecular genetic studies (Selkoe and Toonen 2006) – have been developed only for the Wood Stork (WS). We have attempted cross species amplification of WS markers, using Painted Stork DNA extracted from shed feathers, as a first step in addressing some of the above-mentioned problems (Sharma et al. 2014, 2017b), a summary of which is provided in the table below. The role of Indian zoos in facilitating such work, particularly in obtaining tissue samples, has also been discussed (Sharma et al. 2017a).

Eleven microsatellite loci of Wood Stork whose primers were tested for amplification of homologous regions in Painted Stork DNA obtained from shed feathers at Delhi Zoo (Source: Sharma et al. 2014, 2017b)

Locus	Primer sequence (5′–3′)	Number of alleles	Allele size range (bp)	Amplification success (%)
WSμ03	F-AGAAGCCAAATTGATTAGA R-ACAAAGTTGCGGAGAA	1	166	92.75
WSμ08	F-TGTCTTTCCAGGTAGTTTT R-TACAACTGTTCGTGCTTT	1	180	43.48
WSμ09[a]	F-GGTAACAGCGAGTTGGAT R-TAATGCCAATAAGTGCTTAG	2	270–273	90.83
WSμ13	F-AGGGCTCATCAATAGTGT R-GTTTGCCCACTGTGTCAACT	5	222–230	24.64
WSμ17[a]	F-GGCAAGCTGTTATACTAAT R-GTTTTTCATATACTAACTGG	4	236–246	80.73
WSμ18[a]	F-CATATACTAACTGGGTTTAATC R-GTTGTTCTGCGTTATTC	4	277–287	86.40
WSμ20	F-CGGGCCTTTATCTATC R-ACAGTACCAAACCATTCA	1	141	88.41
WSμ23[a]	F-TTTTGGTGGGATTCATA R-ATAAAAGGGTTAGAAAGACT	5	130–146	89.91
WSμ24	F-GTAAGGCATGAGAGACTAAG R-GTGTGATTTAGGATTGTT	1	237	49.27

[a] Polymorphic loci of high amplification success rate (> 80%) which were used for genetic diversity analysis

References

Abraham, S. (1973) The Kanjirankulam breeding bird sanctuary in the Ramnad district of Tamil Nadu. *Journal of the Bombay Natural History Society* 70: 549–552.

Allendorf, F.W., Leary, R.F., Spruell, P. and Wenburg, J.K. (2001) The problems with hybrids: setting conservation guidelines. *Trends Ecol. Evol* 16: 613–622. https://doi.org/10.1016/S0169-5347(01)02290-X

Ali, S. (1953) The Keoladeo Ghana of Bharatpur (Rajasthan). *Journal of the Bombay Natural History Society* 51: 531–536.

Ali, S. and Ripley, S. D. (1987) *Handbook of the Birds of India and Pakistan Together with Those of Bangladesh, Nepal, Bhutan and Sri Lanka*. New Delhi: Oxford University Press.

Ali, S. and Vijayan, V. S. (1983) Hydrobiological (ecological) research at Keoladeo National Park, Bharatpur (first interim report). Mumbai: Bombay Natural History Society.

Andersson, M. (1994) *Sexual Selection*. Princeton: Princeton University Press.

Barman, P. D., Sharma, D. K., Cockrem, J. F., Malakar, M. and Melvin, T. (2020) Saving the Greater Adjutant Stork by changing perceptions and linking to Assamese traditions in India. *Ethnobiology Letters* 11(2): 20–29. https://doi.org/10.14237/ebl.11.2.2020.1648

Baveja, P., Tang, Q., Lee, J.G.H. and Rheindt, F.E. (2019) Impact of genomic leakage on the conservation of the endangered Milky Stork. *Biological Conservation* 229: 59–66. https://doi.org/10.1016/j.biocon.2018.11.009

Beerens, J. M., Gawlik, D. E., Herring, G. and Cook., M. I. (2011) Dynamic habitat selection by two wading bird species with divergent foraging strategies in a seasonally fluctuating wetland. *Auk* 128: 651–662. https://doi.org/10.1525/auk.2011.10165

Betham, R. M. (1904) Notes on bird's nesting from Poona. *Journal of the Bombay Natural History Society* 15: 709–712.

Bhat, H. R., Jacob, P. G. and Jamgaonkar, A. V. (1991) Observations on a breeding colony of Painted Stork *Mycteria leucocephala* (Pennant) in Anantpur district, Andhra Pradesh. *Journal of the Bombay Natural History Society* 88: 443–445.

Belaire, J. A., Westphal, L. M., Whelan, C. J. and Minor, E. S. (2015) Urban residents' perceptions of birds in the neighborhood: Biodiversity, cultural ecosystem services, and disservices. *Condor* 117: 192–202. http://dx.doi.org/10.1650/Condor-14-128.

Bibby, C., Burgess, N. D. and Hill, D. A. (1992) *Bird Census Techniques*. Oxford: Academic Press.

Bildstein, K. L. (1987) Energetic consequences of sexual size dimorphism in White Ibises (*Eudocimus albus*). *Auk* 104: 771–775. https://doi.org/10.1093/auk/104.4.771

Bildstein, K. L. (1993) *White Ibis: Wetland Wanderer*. Washington, DC: Smithsonian Institution Press.

Biju Kumar, A. (2000). Exotic fishes and freshwater fish diversity. *Zoosprint Journal* 15: 363–367. https://doi.org/10.11609/jott.zpj.15.11.363-7

Bino, G., Seinfeld., C. and Kingsford, R. T. (2014) Maximizing colonial waterbirds' breeding events using identified ecological thresholds and environmental flow management. *Ecological Applications* 24: 142–157. https://doi.org/10.1890/13-0202.1

BirdLife International (2023a) Species factsheet: *Mycteria leucocephala*. Accessed at: http://datazone.birdlife.org/species/factsheet/painted-stork-mycteria-leucocephala (6 October 2022).

BirdLife International (2023b)Species factsheet: *Mycteria cinerea*. Accessed at: http://datazone.birdlife.org/species/factsheet/milky-stork-mycteria-cinerea (6 October 2022)

BirdLife International (2023c) Important Bird Areas factsheet: Kunthangulam Bird Sanctuary. Accessed at: http://www.birdlife.org (5 April 2022).

Birds of the World (2023) Storks. Cornell Lab of Ornithology. Accessed at: https://birdsoftheworld.org/bow/species/wonsto1/cur/introduction (26 January 2022).

Blanford, W. T. (1898) *The Fauna of British India, Including Ceylon and Burma. Birds, Vol. IV*. London: Taylor and Francis.

Bosch, M. (1996) Sexual size dimorphism and determination of sex in yellow-legged gulls. *Journal of Field Ornithology* 67: 534–541.

Bowman, R. (1987) Size dimorphism in mated pairs of American Kestrels. *Wilson Bulletin* 99: 465–467.

British Trust for Ornithology (2022) Heronries surveys. British Trust for Ornithology, Nunnery, Thetford. Accessed at: http://www.bto.org/survey/heron.htm. (26 January 2022).

Brown, C. R. and Brown, M. B. (2001) Avian coloniality, progress and problems. In: Nolan, V. Jr et al. (eds) *Current Ornithology*, vol. 16. New York: Kluwer Academic/Plenum Publishers. https://doi.org/10.1007/978-1-4615-1211-0_1

Burger, J. (1981) A model for the evolution of mixed-species colonies of Ciconiiformes. *Quarterly Review of Biology* 56: 143–167. https://doi.org/10.1086/412176

Campbell, I. C., Poole, C., Giesen, W. and Valbo-Jorgensen, J. (2006) Species diversity and ecology of Tonle Sap Great Lake, Cambodia. *Aquatic Science* 68: 355–373. https://doi.org/10.1007/s00027-006-0855-0

Central Zoo Authority (2023) Who we are. https://cza.nic.in/page/en/introduction Accessed 6 October 2022.

Chace, J. F. and Walsh, J. J. (2006) Urban effects on native avifauna: a review. *Landscape Urban Planning* 74: 46–69. https://doi.org/10.1016/j.landurbplan.2004.08.007

Chardine, J. W. and Morris, J. W. (1989) Sexual size dimorphism and assortative mating in the Brown Noddy. *Condor* 91: 868–874. https://doi.org/10.2307/1368071

Chouhan, M. (2006) Contending water uses: Biodiversity vs irrigation. Case of Keoladeo National Park. *Economic and Political Weekly*, 18 February 2006: 575–577.

Coulson, J. . and Dixon, F. (1979) Colonial breeding in seabirds. In: Larwood, G. and Rosen, B. R. (eds) *Biology and Systematics of Colonial Organisms*. London: Academic Press.

Coulter, M. C. (1986) Assortative mating and sexual dimorphism in the Common Tern. *Wilson Bulletin* 98: 93–100.

Coulter, M. C. and Bryan Jr, A. L. (1993) Foraging ecology of wood storks (*Mycteria americana*) in east-central Georgia. I. Characteristics of foraging sites. *Colonial Waterbirds* 16: 59–70. https://doi.org/10.2307/1521557

Cornell Lab of Ornithology (2023) NestWatch. Cornell Lab of Ornithology. Accessed at: https://nestwatch.org/about/overview/ (26 January 2022)

Collister, D. M. and Wilson, S. (2007) Contributions of weather and predation to reduced breeding success in a threatened northern loggerhead shrike population. *Avian Conservation and Ecology* 2: 11.

Cox, W. A., Thompson., F. R. and Reidy, J. L. (2013) The effects of temperature on nest predation by mammals, birds, and snakes. *Auk* 130: 784–790. https://doi.org/10.1525/auk.2013.13033

Crick, H. Q. P (2004) The impact of climate change on birds. *Ibis* 146(1): 48–56. https://doi.org/10.1111/j.1474–919X.2004.00327.x

Crick, H. Q. P. and Sparks, T. H.(1999) Climate change related to egg-laying trends. *Nature* 399: 423–424. https://doi.org/10.1038/20839

Czech, B. and Krausman, P. R. (1997) Distribution and causation of species endangerment in the United States. *Science* 277: 1116–1117. https://doi.org/10.1126/science.277.5329.1116

Danchin, E. and Wagner, R. H. (1997) The evolution of coloniality: The emergence of new perspectives. *Trends in Ecology and Evolution* 12: 342–347. https://doi.org/10.1016/s0169-5347(97)01124-5

Datta, T. and Pal, B. C. (1993) The effect of human interference on the nesting of the Openbill Stork *Anastomus oscitans* at the Raiganj wildlife sanctuary, India. *Biological Conservation* 64: 149–154. https://doi.org/10.1016/0006-3207(93)90651-g

Datta, T. and Pal, B. C. (1995) Polygyny in the Asian Openbill (*Anastomus oscitans*). *Auk* 112: 257–260. https://doi.org/10.2307/4088788

Davidson, N. C., Townshend, D. J., Pienkowski, M. W. and Speakman, J. R. (1986) Why do curlews *Numenius* have decurved bills? *Bird Study* 33: 61–69. https://doi.org/10.1080/00063658609476896

Dave, K. N. (1985) *Birds in Sanskrit Literature*. New Delhi: Motilal Banarsidass.

Dawson, R. D. and Bortolotti, G. R. (2000) Reproductive success of American Kestrels: The role of prey abundance and weather. *Condor* 102: 814–822. https://doi.org/10.1093/condor/102.4.814

Delacour, J. (1947) *Birds of Malaysia*. New York: Macmillan.

del Hoyo, J., Elliott, A. and Sargatal, J. (eds) (1992) *Handbook of the Birds of the World, Vol. 1*. Barcelona: Lynx Edicions.

Desai, J. H. (1971) Feeding ecology and nesting of painted stork *Ibis leucocephalus* at Delhi Zoo. *International Zoo Yearbook* 11: 208–215. https://doi.org/10.1111/j.1748-1090.1971.tb01908.x

Desai, J. H., Shah, R. V. and Menon, G. K. (1974) Diet and food requirement of painted stork at the breeding colony in the Delhi zoological park. *Pavo* 12: 13–23.

Desai, J. H., Menon, G. K. and Shah, R. V. (1977) Studies on the reproductive pattern of the Painted Stork, *Ibis leucocephalus*, Pennant. *Pavo* 15: 1–32.

Dinsmore, S. J., White, G. C. and Knopf, F. L. (2002) Advanced techniques for modeling avian nest survival. *Ecology* 83: 3476–3488. https://doi.org/10.1890/0012-9658(2002)083[3476:atfman]2.0.co;2

Dreitz, V. J., Conrey, R. Y. and Skagen, S. K. (2012) Drought and cooler temperatures are associated with higher nest survival in Mountain Plovers. *Avian Conservation and Ecology* 7: 6. https://doi.org/10.5751/ace-00519-070106

Djerdali, S., Tortosa, F. S., Hillstrom, L. and Doumandji, S. (2008) Food supply and external cues limit the clutch size and hatchability in the White Stork *Ciconia ciconia. Acta Ornithologica* 43: 145–150.

Fairbairn., D. J. (1997) Allometry for sexual size dimorphism: Pattern and process in the coevolution of body size in males and females. *Annual Review of Ecology and Systematics* 28: 659–687. https://doi.org/10.1146/annurev.ecolsys.28.1.659

Frederick, P. (1985) Intraspecific food piracy in White Ibises. *Journal of Field Ornithology* 56: 413–414.

Frederick, P. C. and Collopy, M. W. (1989) Nesting success of five Ciconiiform species in relation to water conditions in the Florida Everglades. *Auk* 106: 625–634.

Frederick, P. C. and Loftus, W. F. (1993) Responses of marsh fishes and breeding wading birds to low temperatures. *Estuaries* 16: 216–222. https://doi.org/10.2307/1352492

Frederick, P. C. and Spalding, M. G. (1994) Factors affecting reproductive success of wading birds (Ciconiiformes) in the Everglades ecosystem. In S. Davis and J. Ogden (eds) *Everglades: The Eco-system and its Restoration.* Florida: St. Lucie Press.

Frederick, P. and Meyer, K. D. (2008) Longevity and size of wood stork (*Mycteria americana*) colonies in Florida as guides for an effective monitoring strategy in the southeastern United States. *Waterbirds* 31: 12–18. https://doi.org/10.1675/1524-4695(2008)31[12:lasows]2.0.co;2

Frederick, P., Gawlik, D. E., Ogden, J. C., Cook, M. I. and Lusk, M. (2009) The White Ibis and Wood Stork as indicators for restoration of the Everglades ecosystem. *Ecological Indicators* 9: 83–95. https://doi.org/10.1016/j.ecolind.2008.10.012

Fox, G. A. and Weseloh, D. V. (1987) Colonial waterbirds as bio-indicators of environmental contamination in the Great Lakes. In A. W. Diamond and F. L. Fillion (eds) *The Value of Birds.* Cambridge: International Council for Bird Protection.

Fujioka, M. (1986) Two cases of bigyny in the Cattle Egret *Bubulcus ibis. Ibis* 128: 419–422. https://doi.org/10.1111/j.1474-919x.1986.tb02692.x

Gaston, A. J. (1994) Some comments on the 'revival' of Sultanpur lake. *Oriental Bird Club Bulletin* 20: 49–50.

Gopal, B. and Sah, M. (1993) Conservation and management of rivers in India: Case study of the River Yamuna. *Environmental Conservation* 20: 243–254. https://doi.org/10.1017/s0376892900023031

Green, A. J. and Elmberg, J. (2013). Ecosystem services provided by waterbirds. *Biology Review.* doi: 10.1111/brv.12045.

Greenwood, J. G. (2003) Measuring sexual size dimorphism in birds. *Ibis* 145: 124–126. https://doi.org/10.1046/j.1474-919x.2003.00175.x

Greenwood, J. J. D. (2007) Citizens, science and bird conservation. *Journal of Ornithology* 148: 77–124. https://doi.org/10.1007/s10336–007–0239–9

Grimmett, R., Inskipp, C. and Inskipp, T. (1998) *Birds of the Indian Subcontinent.* New Delhi: Oxford University Press.

Hamilton, W. D. (1971) Geometry for the selfish herd. *Journal of Theoretical Biology* 31: 12–45. https://doi.org/10.1016/0022-5193(71)90189-5

Hancock, J. A., Kushlan, J. A. and Kahl, M. P. (1992) *Storks, Ibises and Spoonbills of the World.* London: Academic Press.

Hanson, A. and Kerekes, J. (2006) (eds). *Proceedings of the 4th Conference of the Working Group on Aquatic Birds of the International Society of Limnology* (SIL), *'Limnology and Aquatic Waterbirds 2003'* Sackville, New Brunswick, Canada, August 3–7, 2003. (Developments in Hydrobiology). *Hydrobiologia.* https://doi.org/10.1007/978-1-4020-5556-0

Harrington, R., Woiwod, I. P. and Sparks, T. H. (1999) Climate change and trophic interactions. *Trends in Ecology & Evolution* 14: 146–150. https://doi.org/10.1016/s0169-5347(99)01604-3

Helfenstein, F., Danchin, E. and Wagner, R. H. (2004) Assortative mating and sexual size dimorphism in Black-legged Kittiwakes. *Waterbirds* 27: 350–354. https://doi.org/10.1675/1524-4695(2004)027[0350:amassd]2.0.co;2

Henry, G. M. (1971) A Guide to the Birds of Ceylon, 2nd edn. Kandy, Sri Lanka: K. V. De Silva and Sons.

Hill, D. O. (2014) In Memoriam: M. Philip Kahl, 1934–2012. *Auk* 131(4): 782–786. https://doi.org/10.1642/auk-14-178.1

Hulscher, J. B. (1996) Food and feeding behaviour. In: Goss-Custard J. D. (ed.) *The Oystercatcher: From Individuals to Populations.* Oxford: Oxford University Press.

Hume, A. O. and Oates, E. W. (1890) *The Nests and Eggs of Indian birds, Vol III.* London: Taylor and Francis.

Hutchinson, G. E. (1950) The biogeochemistry of vertebrate excretion. *Bulletin of the American Museum of Natural History* 96: 554.

Indrawan, M., Lawler, W., Widodo W. and Sutandi (1993) Notes on the feeding behaviour of Milky Storks *Mycteria cinerea* at the coast of Indramayu, west Java. *Forktail* 8: 143–144.

Iqbal, M. and Ridwan, A. H. (2008) Local people's perspective for Milky Stork: A case from South Sumatra, Indonesia. *Journal of Wetlands Ecology* 1: 7–8. https://doi.org/10.3126/jowe.v1i1.1571

Iqbal, M., Takari, F., Mulyono, H. and Rasam (2009) A note on the breeding success of Milky Stork *Mycteria cinerea* in 2008, South Sumatra province, Indonesia and more on its diet. *BirdingASIA* 11: 73–74.

Ishtiaq, F., Rahmani, A. R., Javed, S. and Coulter. (2004) Nest site characteristics of Black-necked Stork (*Ephippiorhynchus asiaticus*) and White-necked Stork (*Ciconia episcopus*) in Keoladeo National Park, Bharatpur, India. *Journal of the Bombay Natural History Society* 101: 90–95.

Ishtiaq, F., Javed, S., Coulter, M.C. and Rahmani, A. R.(2010) Resource partitioning in three sympatric species of storks in Keoladeo National Park, India. *Waterbirds* 33(1): 41–49. https://doi.org/10.1675/063.033.0105

Jamgaonkar, A.V., Jacob, P.G., Rajagopal, S.R. and Bhat, H.R. (1994) Records of new breeding colonies of Painted Stork *Mycteria leucocephala* in Karnataka. *Pavo* 32: 59–62.

Jerdon, T. C. (1864) *The Birds of India* (Vol. III). Calcutta: George Wyman and Co

Jhingran, V. G. (1982) *Fish and Fisheries of India.* Delhi: Hindustan Publishing Corporation.

Jovani, R. and Tella, J. L. (2007) Fractal bird nest distribution produces scale-free colony sizes. *Proceedings of the Royal Society B: Biological Sciences* 274: 2465–2469. doi:10.1098/rspb.2007.0527

Kahl, M. P. (1963) Thermoregulation in the Wood Stork, with special reference to the role of the legs. *Physiological Zoology* 36: 141–151. https://doi.org/10.1086/physzool.36.2.30155437

Kahl, M. P. (1970) Observations on the breeding of storks in India and Ceylon. *Journal of the Bombay Natural History Society* 67: 453–461.

Kahl, M. P. (1971a) Spread-wing postures and their possible functions in the Ciconiidae. *Auk* 88: 715–722. https://doi.org/10.2307/4083833

Kahl, M. P. (1971b) The courtship of storks. *Natural History* 80: 36–45.

Kahl, M. P. (1972a) Comparative ethology of the Ciconiidae. The Wood Storks (genera *Mycteria* and *Ibis*). *Ibis* 114: 15–29. https://doi.org/10.1111/j.1474-919x.1972.tb02586.x

Kahl, M. P. (1972b) Comparative ethology of the Ciconiidae. Part 4. The 'typical' storks (genera *Ciconia*, *Sphenorhynchus*, *Dissoura*, and *Euxenura*). *Zeitschrift für Tierpsychologie* 30: 225–252. https://doi.org/10.1111/j.1439-0310.1972.tb00852.x

Kahl, M. P. (1972c) Comparative ethology of the Ciconiidae. Part 5. The Openbill Storks (genus Anastomus). Journal of Ornithology 113: 121–137. https://doi.org/10.1007/bf01640497

Kahl, M. P. (1972d) A revision of the Family Ciconiidae (Aves). *Journal of Zoology* 167: 451–461. https://doi.org/10.1111/j.1469-7998.1972.tb01736.x

Kahl, M. P. (1973) Comparative ethology of the Ciconiidae. Part 6. The Blacknecked, Saddlebill and Jabiru storks (genera *Xenorhynchus*, *Ephippiorhynchus* and *Jabiru*). *Condor* 75: 17–27. https://doi.org/10.2307/1366532

Kahl, M. P. (1974) Comparative behavior and ecology of Asian storks. In: P. Oehser, *National Geographic Society Research Reports: 1967 Projects*. Washington: National Geographic Society.

Kahl, M. P. and Peacock, L. J. (1963) The bill-snap reflex: A feeding mechanism in the American Wood Stork. *Nature* 199: 505–506. https://doi.org/10.1038/199505a0

Kalam, A. and Urfi, A. J. (2008) Foraging behaviour and prey size of the painted stork (*Mycteria leucocephala*). *Journal of Zoology* 274: 198–204. https://doi.org/10.1111/j.1469-7998.2007.00374.x

Kaluthota, C. D., Gamage, S. N. and Kaluthota, U. L. S. (2005) Breeding status of the Painted Stork *Mycteria leucocephala* in the Kumana-Villu of the Yala East National Park. Forestry and Environment Symposium, Colombo, Sri Lanka.

Kazmierczak, K. (2000) *A Field Guide to the Birds of India, Sri Lanka, Pakistan, Nepal, Bhutan, Bangladesh and the Maldives*. New Delhi: Om Book Service.

Kerekes, J. J. and Pollard, B. (1994) (eds) Aquatic birds in the trophic web of lakes. Symposium proceedings, Developments in Hydrobiology, Sackville, New Brunswick, Canada. 19–22 August, 1991 *Hydrobiologia* 96(279/280): 524. https://doi.org/10.1007/978-94-011-1128-7

Khan, M. F. and Panikkar, P. (2009) Assessment of impacts of invasive fishes on the food web structure and ecosystem properties of a tropical reservoir in India. *Ecological Modelling* 220: 2281–2290. https://doi.org/10.1016/j.ecolmodel.2009.05.020

Kharitonov, S. P. and Siegel-Causey, D. (1988) Colony formation in seabirds. *Current Ornithology* 5: 223–272.

Kleindorfer, S., O'Connor, J.A., Dudaniec, R.Y., Myers, S.A., Robertson, J. and Sulloway, F.J. (2014) Species collapse via hybridization in Darwin's tree finches. *The American Naturalist* 183: 325–341. https://doi.org/10.1086/674899

Krebs, J. R. and Davies, N. B. (eds) (1984) *Behavioural Ecology: An Evolutionary Approach*. London: Blackwell Scientific Publications.

Kushlan, J. A. (1977) Sexual dimorphism in the White Ibis. *Wilson Bulletin* 89: 92–98.

Kushlan, J. A. (1978) In: A. Sprunt, J. C. Ogden and S. Winckler (eds) *Feeding Ecology of Wading Birds*. National Audubon Society Research Report No. 7. New York: National Audubon Society.

Kushlan, J. A. (2006) In memoriam: James A. Hancock, 1921–2004. *Auk* 123(1): 278–279. https://doi.org/10.1093/auk/123.1.278

Lack, D. (1968) *Ecological Adaptations for Breeding in Birds*. London: Chapman and Hall.

Li, Z. W. D., Yatim, S. H., Howes, J. and Ilias, R. (2006) Status overview and recommendations for the conservation of Milky Stork *Mycteria cinerea* in Malaysia: Final report of the 2004/2006 Milky Stork field surveys in the Matang Mangrove Forest, Perak. Kuala Lumpur, Malaysia, Wetlands International and the Department of Wildlife and National Parks, Peninsular Malaysia.

Loo, Y. Y., Billa, L. and A. Singh. (2014) Effect of climate change on seasonal monsoon in Asia and its impact on the variability of monsoon rainfall in Southeast Asia. *Geoscience Frontiers* 6: 1–7. https://doi.org/10.1016/j.gsf.2014.02.009

McClure, C. J., Korte, A. C., Heath, J. A. and Barber, J. R. (2015) Pavement and riparian forest shape the bird community along an urban river corridor. *Global Ecology Conservation* 4: 291–310. https://doi.org/10.1016/j.gecco. 2015.07.004

McCreedy, C. and van Riper, C., III (2015) Drought-caused delay in nesting of Sonoran desert birds and its facilitation of parasite- and predator-mediated variation in reproductive success. *Auk* 132: 235–247. https://doi.org/10.1642/auk-13-253.1

Maheswaran, G. and Rahmani, A. R. (2002) Foraging behaviour and feeding success of the black-necked stork (*Ephippiorhynchus asiaticus*) in Dudwa national park, Uttar Pradesh, India. *Journal of Zoology* 258: 189–195. https://doi.org/10.1017/s0952836902001309

Mahendiran, M. and Urfi, A. J. (2010) Foraging patterns and kleptoparasitism among three sympatric cormorants (*Phalacrocorax* spp.) from the Delhi region, North India. *Hydrobiologia* 638: 21–28. https://doi.org/10.1007/s10750-009-0002-8

Mahendiran, M., Parthiban, M. & Azeez, P. A. (2022) Signals of local bioclimate-driven ecomorphological changes in wild birds. *Scientific Reports* 12: 15815. https://doi.org/10.1038/s41598-022-20041-w

Mandel, J. T. and Bildstein, K. L. (2007) Turkey vultures use anthropogenic thermals to extend their daily feeding activity period. *Wilson Journal of Ornithology* 119: 102–105. https://doi.org/10.1676/05-154.1

Manu, K. and Jolly, S. (2000) *Pelicans and People: The Two-Tier Village of Kokkare Bellur, Karnataka, India. Community based conservation in South Asia: Case study no. 4*. New Delhi and London: Kalpavriksh and International Institute of Environment and Development.

Marzluff, J. M., Bowman, R. and Donnelly, R. (eds) (2001) Avian ecology and conservation in an urbanizing world. Boston: Kluwer Academic Publishers.

Mayfield, H. F. (1961) Nesting success calculated from exposure. *Wilson Bulletin* 73: 255–261.

Mayfield, H. F. (1975) Suggestions for calculating nest success. *Wilson Bulletin* 87: 456–466.

Meganathan, T. and Urfi, A. J. (2009) Inter-colony variations in nesting ecology of Painted Stork (*Mycteria leucocephala*) in the Delhi Zoo (North India). *Waterbirds* 32: 352–356. https://doi.org/10.1675/063.032.0216

Melliger, R. L., Braschler, B., Rusterholz, H. P. and Baur, B. (2018) Diverse effects of degree of urbanisation and forest size on species richness and functional diversity of plants and ground surface-active ants and spiders. *PLoS One* 13: 15.

Muralidharan, S., Jaya Kumar, S. and Dhananjayan, V. (2016) Status of pesticide contamination in birds in India. In A. R. Rahmani, Islam, M. Z. and Kasambe, R. M. (eds) *Important Bird and Biodiversity Areas in India: Priority Sites for Conservation* (Revised and updated). Mumbai: Bombay Natural History Society, Indian Bird Conservation Network, Royal Society for the Protection of Birds and BirdLife International (UK).

Nagulu, V. and Rao, J. V. R. (1983) Survey of South Indian pelicanries. *Journal of the Bombay Natural History Society* 80: 141–143.

Naoroji, R. (1990) Predation by *Aquila* eagles on nestling storks and herons in Keoladeo Ghana National Park, Bharathpur. *Journal of the Bombay Natural History Society* 87: 37–46.

Neginhal, S. G. (1977) Discovery of a pelicanry in Karnataka. *Journal of the Bombay Natural History Society* 74: 169–170.

Nice, M. M. (1962) Development of behavior in precocial birds. *Transactions of the Linnaean Society of New York* 8: 1–211.

Niemela, J. (1999) Ecology and urban planning. *Biodiversity Conservation* 8: 119–131. https://doi.org/10.1023/a:1008817325994

Ogden, J. C., Kushlan, J. A. and Tilmant, J. T. (1976) Prey selectivity by the wood stork. *Condor* 78: 324–330.

Owens, N. W. (1984) Why do Curlews have curved beaks? *Bird Study* 31: 230–231. https://doi.org/10.1080/00063658609476896

Padmanabhan, P. and Yom-Tov, Y. (2000) Breeding season and clutch size of Indian passerines. *Ibis* 142: 75–81. https://doi.org/10.1111/j.1474-919x.2000.tb07686.x

Pande, S. (2006) Bhadalwadi Tank, a refuge for Painted Storks. *Hornbill* April–June: 11–15.

Parasharya, B. M. and Naik, R. M. (1990) Ciconiiform birds breeding in Bhavnagar city, Gujarat. In J. C. Daniel and J. C. Serrao (eds) *Conservation in Developing Countries: Problems and Prospects*. Mumbai: Bombay Natural History Society.

Pattanaik, C., Prasad, S. N., Murthy, E. N. and Reddy, C. S. (2008) Conservation of Painted Stork habitats in Andhra Pradesh. *Current Science* 95: 1001.

Paulraj, S. (1988) Impact of guano deposition in Vedanthangal Water-bird Sanctuary (Chengalpattu district, Tamil Nadu). *Journal of the Bombay Natural History Society* 85: 319–324.

Paulraj, S. and Kondas, S. (1987) Two centuries of protection of a waterbird colony: How it may benefit the protectors. *Environmental Conservation* 14: 363–364. https://doi.org/10.1017/s0376892900016891

Pearce-Higgins, J. W. and Green, R. E. (2014) *Birds and Climate Change*. Cambridge: Cambridge University Press. https://doi.org/10.1017/CBO9781139047791

Pennant, T. (1790) *Indian Zoology*. London: Henry Hughs for Robert Faulder.

Pennycuick, C. J. (1972) Soaring behaviour and performance of some East African birds, observed from a motor glider. *Ibis.* 114: 178–218.

Perrins, C. M. and Birkhead, T. (1983) *Avian Ecology.* London: Blackie.

Poole, C. (1994) Sultanpur lake revived. *Oriental Bird Club Bulletin* 19: 15.

Rahmani, A. R. (2012) *Threatened Birds of India and Their Conservation Requirements.* Mumbai: Bombay Natural History Society.

Rahmani, A. R., Islam, M .Z. and Kasambe, R. M. (2016) *Important Bird and Biodiversity Areas in India: Priority Sites for Conservation* (Revised and updated). Mumbai: Bombay Natural History Society, Indian Bird Conservation Network, RSPB and BirdLife International (UK).

Rasmussen, P. C. and Anderton J. C. (2005) *Birds of South Asia: The Ripley Guide, Vols 1 and 2.* Washington DC: Lynx Edicions.

Rawat, M., Moturi, M. C. Z. and Subramaniam, V. (2003) Inventory compilation and distribution of heavy metals in wastewater from small-scale industrial areas of Delhi. *Journal of Environmental Monitoring* 5: 906–912. https://doi.org/10.1039/b306628b

Rehman, S., Tiwary, N. K. and Urfi, A. J. (2021) Conservation monitoring of a polluted urban river: An occupancy modeling study of birds in the Yamuna of Delhi. *Urban Ecosystems* (2021). https://doi.org/10.1007/s11252–021–01127–1

Rhenius, C. E. (1907) Pelicans breeding in India. *Journal of the Bombay Natural History Society* 17: 806–807.

Ripley, S. D. and Beehler, B. M. (1990) Patterns of speciation in Indian birds. *Journal of Biogeography* 17: 639–648. https://doi.org/10.2307/2845145

Rising, J. D. and Somers, K. M. (1989) The measurement of overall body size in birds. *Auk* 106: 666–674.

Roberts, T. J. (1991) *The Birds of Pakistan.* 2 vols. Karachi: Oxford University Press.

Robson, C. (2000) *A Field Guide to the Birds of South-East Asia.* London: New Holland.

Rodgers, J. A., Jr. (1990) Breeding chronology and clutch information for the Wood Stork from museum collections. *Journal of Field Ornithology* 61: 47–53.

Rodgers, J. A., Jr., Wenner, A. S. and Schwikert, S. T. (1988) The use and function of green nest material by wood storks. *Wilson Bulletin* 100: 411–423.

Rhymer, J.M. and Simberloff, D. (1996) Extinction by hybridization and introgression. *Annu. Rev. Ecol. Syst.* 27: 83–109. https://doi.org/10.1146/annurev.ecolsys.27.1.83

Sandercock, B. K. (1998) Assortative mating and sexual size dimorphism in Western and Semipalmated Sandpipers. *Auk* 115: 786–791. https://doi.org/10.2307/4089430

Sankhala, K. (1990) *Gardens of God. The Waterbird Sanctuary at Bharatpur.* New Delhi: Vikas Publishing House.

Savard, J. P. L., Clergeau, P. and Mennechez, G. (2000) Biodiversity concepts and urban ecosystems. *Landscape and Urban Planning.* 48: 131–142. https://doi.org/10.1016/s0169-2046(00)00037-2

Schulte-Hostedde, A. I. and Millar, J. S. (2000) Measuring sexual size dimorphism in the Yellow-pine Chipmunk (*Tamias amoenus*). *Canadian Journal of Zoology* 78: 728–733. https://doi.org/10.1139/z00-005

Selander, R. K. (1966) Sexual dimorphism and differential niche utilization in birds. *Condor* 68: 113–151. https://doi.org/10.2307/1365712

Selander, R. K. (1972) *Sexual selection and dimorphism in birds.* In: B. Campbell (ed.) *Sexual Selection and the Descent of Man 1871–1971.* Chicago: Aldine.

Selkoe, K. A. and Toonen, R. J. (2006) Microsatellites for ecologists: A practical guide to using and evaluating microsatellite markers. *Ecology Letters* 9: 615–629. https://doi.org/10.1111/j.1461-0248.2006.00889.x

Shah, R. V., Menon, G. K., Desai, J. H and Jani, M. B. (1977a) Feather loss from capital tracts of Painted Storks related to growth and maturity. Part 1. Histophysiological changes and lipoid secretion in the integument. *Journal of Animal Morphology and Physiology* 24: 98–107.

Shah, R.V., Menon, G. K., Jani, M. B. and Desai, J. H. (1977b) Feather loss from capital tract related to growth and maturity of Painted Storks. Part 2. Distribution of glycosaminoglycans and phosphatases in the integument. *Pavo* 15: 155–163.

Sharma, P. K. and Raghavaiah, P. S. (2000) Breeding of Painted Stork at Vedurupattu, Nellore district, Andhra Pradesh: Co-existence of man and wildlife. *Indian Forester* 126: 1147–1149.

Sharma, B. B., Mustafa, M., Sharma, T., Banerjee, B. D. and Urfi, A. J. (2014) Cross-species amplification of microsatellite markers in *Mycteria leucocephala* Pennant 1769: Molted feathers as successful DNA source. *Indian Journal of Experimental Biology* 52 (Oct 2014): 1011–1016. http://nopr.niscair.res.in/handle/123456789/29424

Sharma, B. B., Banerjee, B. D. and Urfi, A. J. (2017a) Indian zoos can contribute towards ornithological research in novel ways. *Current Science* 113(6): 1013. http://www.currentscience.ac.in/Volumes/113/06/1013.pdf

Sharma, B. B., Banerjee, B. D. and Urfi, A. J. (2017b) A preliminary study of cross-amplified microsatellite loci using molted feathers from a near-threatened Painted Stork (*Mycteria leucocephala*) population of north India as a DNA source. *BMC Research Notes.* https://doi.org/10.1186/s13104-017-2932-y

Shine, R. (1989) Ecological causes for the evolution of sexual dimorphism: A review of the evidence. *Quarterly Review of Biology* 64: 419–461. https://doi.org/10.1086/416458

Sibley, C. G. and Ahlquist, J. E. (1990) *Phylogeny and Classification of Birds: A Study in Molecular Evolution.* New Haven: Yale University Press.

Siegel-Causey, D. and Kharitonov, S. P. (1990) The evolution of coloniality. *Current Ornithology* 7: 285–330.

Slatkin, M. (1984) Ecological causes of sexual dimorphism. *Evolution* 38: 622–630. https://doi.org/10.2307/2408711

Slikas, B. (1997) Phylogeny of the avian family Ciconiidae (storks) based on cytochrome b sequences and DNA–DNA hybridization distances. *Molecular Phylogenetics and Evolution* 8: 275–300. https://doi.org/10.1006/mpev.1997.0431

Slikas, B. (1998) Recognizing and testing homology of courtship displays in Storks (Aves: Ciconiiformes: Ciconiidae). *Evolution* 52: 884–893. https://doi.org/10.1111/j.1558-5646.1998.tb03713.x

Smythies, B.E.(1940) *The birds of Burma.* London: Oliver and Boyd.

Starck J. M., Ricklefs R. E. (eds) (1998) *Avian Growth and Development: Evolution Within the Altricial-Precocial Spectrum.* Oxford University Press: New York.

Stephens, D. W. and Krebs, J. R. (1986) *Foraging Theory.* Princeton: Princeton University Press.

Subramanya, S. and Manu, K. (1996) Saving the spot-billed pelican: A successful experiment. *Hornbill* 2: 2–6.

Sundar, K. S. G. (2006) Flock size, density and habitat selection of four large waterbirds species in an agricultural landscape in Uttar Pradesh, India: Implications for management. *Waterbirds* 29: 365–374. https://doi. org/10.1675/1524-4695(2006)29[365:fsdahs]2.0.co;2

Swennen, C. and Marteijn, E. C. L. (1987) Notes on the feeding behaviour of the milky stork (*Mycteria cinerea*). *Forktail* 3: 63–66.

Székely,T., Lislevand ,T. and Figuerola, J. (2007) Sexual size dimorphism in birds. In: D. Fairbairn, W. U. Blanckenhorn and T. Székely (eds) *Sex, Size and Gender Roles*. Oxford: Oxford University Press. https://doi.org/10.1093/acprof:oso/97801992 08784.003.0004

Szostek, K. L., Becker, P. H., Meyer, B. C., Sudmann, S. R. and Zintl, H. (2014) Colony size and not nest density drives reproductive output in the Common Tern *Sterna hirundo*. *Ibis* 156: 48–59. https://doi.org/10.1111/ibi.12116

Taubenböck, H., Wegmann, M., Roth, A., Mehl, H. and Dech, S. (2009) Urbanization in India – Spatiotemporal analysis using remote sensing data. *Computers, Environment and Urban Systems* 33: 179–188. http://dx.doi.org/10.1016/j. compenvurbsys.2008.09.003

Tella, J. L., Hiraldo, F. and Donázar, J. A. (1998) The evolution of coloniality: Does commodity selection explain it all? *Trends Ecology and Evolution* 13: 75–76. https:// doi.org/10.1016/s0169-5347(97)01300-1

Tiwary, N. K. and Urfi, A. J. (2016a) Nest survival in Painted Stork (*Mycteria leucocephala*) colonies of north India: The significance of nest age, annual rainfall and winter temperature. *Waterbirds*. 39(2): 146–155. https://doi.org/10.1675/063. 039.0205

Tiwary, N. K. and Urfi, A. J. (2016b) Spatial variations of bird occupancy in Delhi: The significance of woodland habitat patches in urban centres. *Urban Forestry and Urban Greening* 20: 338–347. https://doi.org/10.1016/j.ufug.2016.10.002

Tiwary, N. K., Sharma, B. B. and Urfi, A. J. (2014). Two new nesting colonies of Painted Stork *Mycteria leucocephala* from northern India. *Indian Birds* 9(4): 85–8.

Todesco, M., Pascual, M.A., Owens, G.L., Ostevik, K.L., Moyers, B.T., Hübner, S., Heredia, S.M., Hahn, M.A., Caseys, C., Bock, D.G., Rieseberg, L.H. (2016) Hybridization and extinction. *Evolutionary Applications* 9: 892–908. https://doi. org/10.1111/eva.12367

Urfi, A. J. (1996) One bird's prison. *New Scientist*: 24 August 1996: 45.

Urfi, A. J. (1997) The significance of Delhi Zoo for wild waterbirds, with special reference to the Painted Stork *Mycteria leucocephala*. *Forktail* 12: 87–97.

Urfi, A. J. (1998) A monsoon delivers storks. *Natural History* 107: 32–39.

Urfi, A. J. (2003a) Record of a nesting colony of Painted Stork *Mycteria leucocephala* at Man-Marodi Island in the Gulf of Kutch. *Journal of the Bombay Natural History Society* 100: 109–110.

Urfi, A. J. (2003b) Breeding ecology of birds. Why do some species nest singly while others are colonial? *Resonance* 8: 22–32.

Urfi, A. J. (2004) *Birds: Beyond Watching*. Hyderabad: Universities Press.

Urfi A. J. (2006a) Biodiversity conservation in an urban landscape. A case study of some important bird areas on the river Yamuna in Delhi (India). In: J. A. McNeely, T. M. McCarthy, A. Smith, L. Olsvig-Whittaker and E. D. Wikramanayake (eds) *Conservation Biology in Asia*. Kathmandu, Nepal: Society for Conservation Biology (Asia Section) and Resources Himalaya Foundation.

Urfi, A. J. (2006b) Mucking around in the Everglades. *Indian Fulbrighter*, November 2006: 6–8.

Urfi, A. J. (2008) *Birds of India – A Literary Anthology*. New Delhi: Oxford University Press.

Urfi, A. J. (2010a) Factors causing nest losses in the Painted Stork (*Mycteria leucocephala*): A review of some Indian studies. *Journal of the Bombay Natural History Society* 107: 55–58.

Urfi, A. J. (2010b) Using heronry birds to monitor urbanization impacts: A case study of Painted Stork *Mycteria leucocephala* nesting in the Delhi Zoo, India. *Ambio* 39: 190–193. https://doi.org/10.1007/s13280-010-0018-3

Urfi, A. J. (2011a) *The Painted Stork: Ecology and Conservation*. New York and Dordrecht: Springer.

Urfi, A. J (2011b) Foraging ecology of the Painted Stork (*Mycteria leucocephala*): A review. *Waterbirds* 34(4): 448–456. https://doi.org/10.1675/063.034.0407

Urfi, A. J. (2012) Birdwatchers, Middle Class and the 'Bharat-India' Divide. Perspectives from Recent Bird Writings. *Economic and Political Weekly* 47(42), October 20: 2349–8846. https://www.epw.in/journal/2012/42/commentary/birdwatchers-middle-class-and-bharat-india-divide.html

Urfi, A.J. (2016) Animal Locomotion in Different Mediums. The Adaptations of Wetland Organisms. *Resonance.* June 2016: 545–556. https://doi.org/10.1007/s12045-016-0359-8

Urfi, A. J. (2017) Values Associated with Village Tanks and Some Instances of Enviro-Legal Activism for Wetland Conservation. (Ch 29). Pp. 563–573. In: *Wetland Science, Perspectives from South Asia*. Edited by Prusty, B. A.K, Chandra, R & P. A. Azeez. Springer (India). https://doi.org/10.1007/978-81-322-3715-0_29

Urfi, A. J. (2020) Indian heronries need conservation monitoring. *Nature India*. doi:10.1038/nindia.2020.12 (23 January 2020). https://www.natureasia.com/en/nindia/article/10.1038/ nindia.2020.12

Urfi, A. J. (2021a) Painted Stork nesting dangerously low. Is people's protective attitude really a factor at Kokkare Bellur (South India)? Blog British Ornithological Union (#theBOUblog), 5 July 2021. https://bou.org.uk/blog-urfi-painted-stork- nesting-dangerously-low/

Urfi, A. J. (2021b) Forgotten stories of India's environment movement. *Nature India*. doi.org/10.1038/nindia.2021.40 (17 March 2021). EISSN: 1755–3180 https://www.natureasia.com/en/nindia/article/10.1038/nindia.2021.40

Urfi, A. J. and Nareshwar, M. (1998) Interpreting a village pond heronry. *NewsEE (CEE)* 4/3: 5.

Urfi, A. J. and Kalam, A. (2006) Sexual size dimorphism and mating pattern in the Painted Stork (*Mycteria leucocephala*). *Waterbirds* 29: 489–496. https://doi.org/10.1675/1524-4695(2006)29[489:ssdamp]2.0.co;2

Urfi, A. J., Sen, M., Kalam, A. and Meganathan, T. (2005) Counting birds in India: A review of methodologies and trends. *Current Science* 89: 1997–2003.

Urfi, A. J., Meganathan, T. and Kalam, A. (2007) Nesting ecology of Painted Stork (*Mycteria leucocephala*) at Sultanpur National Park (Haryana), India. *Forktail* 23: 150–153.

Verheugt, W. J. M. (1987) Conservation status and action programme for the Milky Stork (*Mycteria cinerea*). *Colonial Waterbirds* 10: 211–220. https://doi.org/10.2307/1521260

Wagner, R. H. (1999) Sexual size dimorphism and assortative mating in Razorbills (*Alca torda*). Auk 116: 542–544. https://doi.org/10.2307/4089388

Wagner, R. H., Danchin, E., Boulinier, T. and Helfenstein, F. (2000) Colonies as byproducts of commodity selection. *Behavioral Ecology* 11: 572–573. https://doi.org/10.1093/beheco/11.5.572

Wanless, S. and Harris, M.P. (1991) Diving patterns of full-grown and juvenile Rock Shags. *Condor* 93: 44–48. https://doi.org/10.2307/1368604

Webb-Peploe, C. G. (1945) Notes on a few birds from the south of the Tinnevelly district. *Journal of the Bombay Natural History Society* 45: 425–426.

Wells, D. R. (1999) *The Birds of the Thai–Malay Peninsula: covering Burma and Thailand south of the eleventh parallel, Peninsular Malaysia and Singapore*, vol. 1, *Non-passerines*. London: Academic Press.

Westneat, D. F., Sherman, P. W. and Morton, M. L. (1990) The ecology and evolution of extra-pair copulation in birds. *Current Ornithology* 7: 331–369.

Wiens, J. A. (1973) Patterns and process in grassland bird communities. *Ecological Monographs* 43: 237–270. https://doi.org/10.2307/1942196

Wilkinson, M. E. (1961) Pelicanry at Kundakulam, Tirunelveli district. *Journal of the Bombay Natural History Society* 58: 514–515.

Wittenberger, J. F. and Hunt, G.L., Jr (1985) The adaptive significance of coloniality in birds. In: D. S. Farner, J. R. King (eds) *Avian Biology*, vol 8. San Diego: Academic Press.

Wolf, D.E., Takebayashi, N. and Rieseberg, L.H. (2001) Predicting the risk of extinction through hybridization. *Conservation Biology* 15: 1039–1053. https://doi.org/10.1046/j.1523-1739.2001.0150041039.x

Yang, G., Xu, J., Wang, Y., Wang, X., Pei, E., Yuan, X., Li, H., Ding, Y. and Wang, Z. (2015) Evaluation of microhabitats for wild birds in a Shanghai urban area park. *Urban Forestry and Urban Greening* 14: 246–254. http://dx.doi.org/10.1016/j.ufug.2015.02.005.

Zipkin E. F., DeWan, A. and Royle, A. (2009) Impacts of forest fragmentation on species richness: A hierarchical approach to community modelling. *Journal of Applied Ecology* 46: 815–822. https://doi.org/10.1111/j.1365–2664.2009.01664.x

Žydelis R. and Kontautas, A. (2008). Piscivorous birds as top predators and fishery competitors in the lagoon ecosystem *Hydrobiologia* 611: 45–54. https://doi.org/10.1007/s10750-008-9460-7

Index

Page numbers in *italics* refer to illustrations.